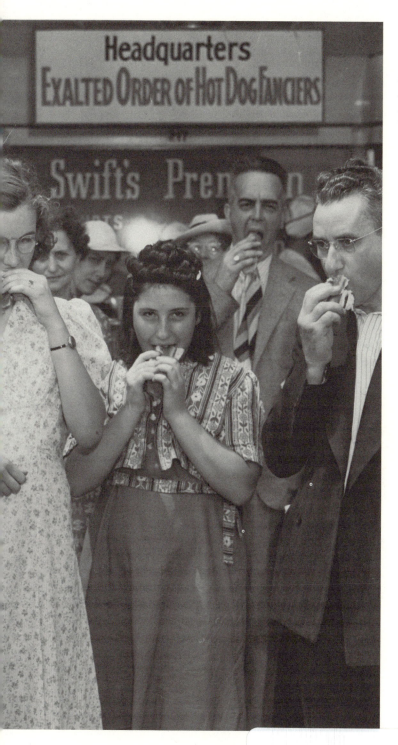

*Putting
Meat on the
American
Table*

D0024287

Putting Meat on the American Table

Taste, Technology, Transformation

Roger Horowitz

THE Johns Hopkins UNIVERSITY PRESS
Baltimore

7-17-06
ww
$19—

The Johns Hopkins University Press
2715 North Charles Street
Baltimore, Maryland 21218-4363
www.press.jhu.edu

Library of Congress Cataloging-in-Publication Data

Horowitz, Roger.
Putting meat on the American table: taste, technology, transformation / Roger Horowitz.
p. cm.
Includes bibliographical references and index.
ISBN 0-8018-8240-0 (hardcover: alk. paper) —
ISBN 0-8018-8241-9 (pbk.: alk. paper)
1. Meat industry and trade—United States—History. I. Title.
HD9415.H67 2006
388.1'76'00973—dc22   2005009016

A catalog record for this book is available from the British Library.

Frontispiece: New York World's Fair (1939–40) Records, Manuscripts and Archives Division, The New York Public Library, Astor, Lenox and Tilden Foundations

To my parents
David H. Horowitz
and Louise Schwartz Horowitz
for all that they have done,
and for all that they have done for me

Contents

Illustrations

x

Illustrations

Preface

This book began with a question: what is the relationship between production and consumption of goods in our society, and how does this relationship work itself out in a product like meat, eaten every day, and for hundreds of years, by millions of Americans? How can such a quotidian case of consumer demand manifest itself in the complex of institutions erected to satisfy that daily hunger?

The inspiration for this inquiry came from editor Robert J. Brugger at the Johns Hopkins University Press who asked if I would consider writing a book on technology and meat consumption. Having completed several studies of labor and business in meatpacking and knowing a great deal about production technology, I was intrigued by the possibility of researching more deeply something I already knew a great deal about, but around a different set of questions. A subsequent conversation with my colleague at Hagley, Phil Scranton, sharpened my analytic approach to such an undertaking.

I also was influenced by dissatisfaction with books already published about meat. Provocative studies by Jeremy Rifkin, Nick Fiddes, and Carol J. Adams took a critical, indeed denunciatory, stance toward meat, viewing meat consumption as essentially a bad thing for people, animals, and the environment. Regardless of the validity of their arguments, the books did not address why so many Americans, for so long and so consistently, have made meat an essential part of their diet.[1]

In this volume, I avoid moral judgment on popular practices. My objective is, above all, to appreciate how the choices to eat meat emerged from the experiences of Americans' lives, and, within the constraints of the world in which they operated, from efforts to create a good life for themselves and their families. I look closely at the dynamics of meat consumption, especially the tangled relationship between consumers and producers, with the relative power of these parties waxing and waning over time and among products.

To organize the study I treat meat as material culture, an artifact with particular physical characteristics appropriated by human beings in certain specific ways. Thinking about meat's natural properties—and the obstacles thereby posed to mass production—brought me back to a remarkable book written shortly after World War II. Sigfried Giedion's

Mechanization Takes Command postulated that the central problem in modern industrial society was the effort to create technologies that could "take command" of the natural world and in so doing subordinate nature to human will. Giedion's insights on the incomplete accomplishments of such technology meshed with my prior research on the meatpacking industry. I had found that hand labor, and struggles between management and labor to control production, persisted in the industry despite all efforts to develop automatic production machinery. The central obstacles to mechanization "taking command" were the perishable nature of meat products and the dilemma of organizing mass production around an item that came in irregular sizes.

Focusing on consumption further emphasized the difficulties created by meat's perishable nature and origin in animals that varied in weight and shape. Hence the book's principal question expanded slightly, to consider not only how consumption and production were related in meat production but also, given meat's irreducible natural qualities, what kinds of social arrangements developed to facilitate its regular delivery to Americans. And further, how did the ways people consume, handle, preserve, and obtain meat reverberate back up the provisioning chain, to producers, the meatpacking companies, and the farmers who actually raised the animals in the first place?

The book opens by establishing meat's material qualities and its entrenched place in the American diet. Subsequent chapters focus on four distinct meat varieties: beef, pork, hot dogs (sausage), and chicken. Such close studies of what meat Americans actually ate, and how those varieties came to be, allow us to avoid the simplistic explanatory schemes that too often inform studies of meat. The book closes with a chapter emphasizing the transformation of meat following World War II around the logic of convenience.

This book is very much a product of the Hagley Museum and Library, where I worked during the entire time it was under preparation. Hagley's remarkable library on American industrial society formed the single most important resource for this research. Its rich collections cover topics from antebellum New York City and Cincinnati to postwar packaging and food consumption patterns. The book also is shaped by the ongoing intellectual agenda of the Center for the History of Business, Technology, and Society (of which I serve as associate director) to link economic and business history with cultural processes, technology, and labor history. Discussions with the Center's director, Phil Scranton, had a major formative impact on my approach, as did Phil's own scholarship on American industrial production. Engagement with and support from others at Hagley, including Michael Nash, Carol Lockman, Marjorie McNinch, Lynn Catanese, Susan Hengel, Jon Williams, and Barbara Hall affected the book as well. Glenn Porter,

Hagley's director through much of this project, provided unusual latitude for a staff member to devote time to research and writing, along with making me appreciate the importance of packaging and design. His successor, George Vogt, continued Hagley's generous support for my scholarly activities.

I benefited greatly from a fellowship at the Lemelson Center for Invention and Innovation at the National Museum of American History, which gave me time to write and made it possible to explore the museum's research collections and (in the evening) draw on the extraordinary resources of the Library of Congress. Lemelson staff member (and former student) Maggie Dennis encouraged and influenced this project from its earliest days, as did Mimi Minnick in the Archives Center. Craig Orr and John Fleckner in the museum's archives enriched my research by acquiring a unique collection of meat industry trade catalogs just prior to my residency—a magnificent welcoming gift! Art Molella, Janet Davidson, Rayna Green, Joyce Bedi, and Steve Lubar also patiently listened to my ideas and helped influence the ultimate product.

Influences came from elsewhere as well. My mother, Louise Horowitz, read a very early draft, and her incisive critique dramatically altered the book's direction for the better. My father, David Horowitz, formulated the book's final title. The scholarship and intellectual dialogues with my good friends at the University of Delaware, Arwen Mohun and Susie Strasser, deeply affected my approach, as did the responses of students in my class "American Eats" who listened to versions of these chapters delivered in lecture form. Other food historians encountered during this project, especially Sydney Watts, Jeffrey Pilcher, Warren Belasco, and Donna Gabaccia, brought their own perspectives "to the table" and in so doing expanded my appreciation for food as a discrete area of study. I learned as well from reactions to papers drawn from this book delivered at the American Food Studies Association, European Business History Association, American Historical Association, Yale University, the University of Delaware, the Hagley Museum and Library, and the Culinary Historians of Washington, D.C. An earlier version of chapter 5 (on chicken) appeared in *Industrializing Organisms* (Routledge, 2003), edited by Philip Scranton and Susan Schrepfer.

*Putting
Meat on the
American
Table*

Chapter One

A Meat-Eating Nation

I like meat. A grilled rib-eye steak with some crisp bread and red wine, pork stir-fried in hot sauce surrounded by vegetables and accompanied by a beer, roast chicken and potatoes for a late-afternoon Sunday dinner, and, yes, early morning corned beef hash under my fried eggs downed with a quart of coffee before a day spent working on the house. Yet I also am appalled by the deep systemic problems of our meat provisioning system: the health hazards of the meats that come through the processing plants, the difficult working conditions of most slaughterhouse employees, the ecological effect of concentrated livestock production on rural areas. And I am troubled by the inefficient use of our agricultural resources to feed animals when so many people in this world remain malnourished.

My personal dilemma is, and has been, shared by many Americans who make meat a central part of their diet even as they are aware of the many problems in the system that provides so much beef, pork, and chicken. Periodically out of this dilemma have come sensational exposés that seek to reveal disturbing features of animal-based foods, from *The Jungle* (1906) to *Fast Food Nation* (2001), from the in-depth newspaper series to the sensational television news program. If these exposés sought to make change by shocking the public—hitting Americans in the stomach, as it were—the long-term patterns of meat consumption show no lessening of our taste for meat. Americans born since World War II eat more meat than at any other time in the country's history. The principal effect of sensational criticisms of the industry has been in the area of regulation, not consumption levels. Regardless of all the demonstrated problems with the U.S. meat supply, meat remains the centerpiece of American meals. We have been, are, and in the future will remain, a meat-eating nation.

Without doubt, eating meat has been an integral part of the American diet since settlement. While frontier regions relied heavily on abundant game, farming-based districts rapidly developed a stable meat economy. The best systematic information we have indicates that average per capita meat consumption topped 150 pounds annually by the nineteenth century. Aside from a dramatic dip during the depths of the Great Depression, Americans continued to eat meat at this level until the 1960s, when consumption increased dramatically to more than 200 pounds per person per year. For most of American history, eating six to eight ounces of meat each day has been a defining feature of our society.

Attaining such high levels of meat consumption has not been easy. It is difficult to turn a living thing into a meal for human beings. Aside from the troubling philosophical question of what right we have as a species to control other animals (a topic I will not address here), animals' bodies resist becoming an expression of our will. To this day the meat industry remains tethered to a natural product, hemmed in and constrained by the special features of its source.

The dilemma of a meat-eating nation is that meat comes in irregular sizes and begins to deteriorate the instant its vessel, the animal, is killed. Developing the apparatus for killing animals and then preserving and disseminating their meat entailed massive capital investment by business organizations, the labors of hundreds of thousands of workers, and the creation of machinery to speed production and distribution. "Making meat," however, was always tightly linked to the lunch wagon and the dinner table, that is, to the way Americans obtained and ate meat. Meat's irreducible natural features led to repeated cycles of technological innovations producing a continuously changing array of meat products. Processing technologies and entrepreneurial initiatives evolved in close conjunction with food consumption practices and with Americans' insistence on obtaining wholesome and nutritious meat. Throughout, standardizing the shape of meat and slowing physical decay have been the Holy Grail of meat purveyors. In search of these ever-receding objectives, innovation has interacted with—at times clashing with, at times transforming—popular tastes for meat.

What Is Meat?

From the standpoint of physical chemistry, meat is animal muscle fiber consisting of long strings of thin, chainlike cells. It has a unique ability to flex forward and back, imparting movement to a body. Two proteins in these cells, myosin and actin, slide past each other when the proper electrical impulses are received from the central nervous system. They then contract or expand the muscle in accordance with these instructions. Muscles that are frequently used develop cross bridges that hold the proteins together and create a more complex molecule, actomyosin, thickening these fibers—and thereby creating tougher meat.

Muscle fibers and blood vessels bundled together by connective tissue constitute the muscles from which meat is drawn. The "toughness" of meat is directly related to the amount of connective tissue in a particular cut. Consequently, large muscles (with less connective tissue) are intrinsically more tender than small ones. Properly distributed fat—

that is, evenly "marbled" in meat—can moderate the toughness of thick muscle fibers. Cooking melts fat, which then oozes between and separates the fibers, acting as an oil that makes it easier to cut through the bundles of cells with a knife or one's teeth.

The practice of fattening animals before slaughter in feeding situations that restrict movement as much as possible can have dramatic effects on meat quality. Heavily used muscle has relatively thicker filaments and more connective tissue. Hence legs are tougher than the upper body in most animals; reducing movement retards muscular development. Fattening before slaughter increases the distribution of fat through the muscles and thereby creates meat that is more tender when cooked.

The methods of an animal's death dramatically affect its flesh as well. The adrenaline of an animal fearing death creates insufficient lactose acid in the flesh and leads to "dark cut" meat, an inferior product that does not cure or keep well. It has been a maxim among meat purveyors that livestock should be calm and unaware of its coming demise so that the meat is not damaged by the "excitement" of a frightened animal.

Protein is meat's great contribution to the human diet. Protein is composed of several amino acids that have to be present in similar quantities to be used efficiently by the body; meat (and other animal products such as eggs and milk) contains complementary amino acids in relatively equal proportions that permit the protein to be used efficiently. While protein can be obtained from vegetable sources, they are more likely to contain highly imbalanced distribution, with our bodies only able to assimilate that made available by the limiting amino acid, the one available in lowest quantity. Diets based on rice and beans draw on foods that complement available amino acids; their sum is greater than their parts! But meat is simple to eat for this necessary nutritional purpose; even a critic such as Francis More Lappe concedes that 4 to 6 ounces of meat contains 100 percent of an adult's daily protein need.

Meat also has been prized as a compact source of minerals and vitamins. It is an excellent source of B vitamins, and pork is an especially strong supplier of thiamine. Meat also "delivers" iron quite efficiently to the human body, along with other important and hard-to-find minerals such as phosphorous and copper.

While the tactile and taste qualities of meat can vary enormously, the proteins, vitamins, and minerals available from meat are comparatively consistent regardless of variety (with the exception of pork's greater thiamine content). Lean round beef has as much protein as fatty loin steak; variations in food value are due to fat content per pound, as more fat means less true muscle fiber. This physiological character of animal flesh thus helps make sense of the wide use of different parts of the animal, for even if not particularly palatable, meat still provides valuable nutrients for human consumption.

Not everything is good about meat, of course. Its fat content is a mixed blessing. The fatty deposits among muscle fibers soften cooked meat and improve its flavor. The chemical changes initiated by heating damage the cell membranes and allow fat and protein molecules to interact, producing chemical reactions that improve the intensity of taste in the meat. Fat, by liquefying meat, also stimulates saliva in the human mouth and makes meat seem juicier than it may actually be. The nutritional contribution of fat, however, is largely as a source of calories. Fatty meat may be more tasty than lean, but as the fat content rises, meat's contribution to nutrition declines. A porterhouse steak may taste good, but it is not as healthy as lean round steak. And with fat comes cholesterol, technically not a fat but a lipid associated with the kind of saturated fat found in meat. Our bodies need some cholesterol, just as we need fat, but the increase in meat consumption after 1960 is associated with (though not definitively the cause of) significant imbalances of fat and cholesterol in the human diet.[1]

Cooking Meat

Cooking is the medium between slaughter and consumption: the moment when animal flesh becomes human food. While cooking is certainly about taste, it also is about chemistry and reflects vernacular knowledge about the characteristics of the meat from different parts of the animal. Cooking seeks to soften the muscle fibers and weaken connective tissue; to retain juices in the meat to lubricate digestion; to stimulate taste through heat sufficient to generate chemical reactions in the meat; and to create a palatable item for the table. Variations in meat texture and composition, while trivial from a nutritional standpoint, can have dramatic effects on the most efficacious preparation methods.

The principal variables influencing appropriate cooking methods are meat fiber composition and fat content. Meat that comes from larger animals has longer, thicker fibers; beef thus is intrinsically tougher than chicken. An animal's more heavily used muscles are lined with connective tissue, as activity stimulates better developed circulatory and support tissues and thicker muscle fibers. The tenderloin, from the middle of an animal's back, is almost always more tender than the leg. Meat sliced "across the grain" is a cut that goes through these fibers; "stringy" describes well-developed fibrous muscles while "velvety" characterizes those with thinner fibers and less connective tissue development.

Fat content, and placement, also matter greatly during cooking.

When heated, fat distributed through the meat helps weaken the connective tissue by conveying heat that turns collagen, one of its components, to soft gelatin. Layered on the outside of meat, fat bastes the exterior and stimulates chemical reactions on the surface that improve taste. Pork fat is particularly well distributed around the muscle fibers, probably influencing the relative softness of the cured forms of this meat in comparison to beef.

Cooking, viewed as chemistry rather than cuisine, attends to these variables by altering methods depending on the flesh's composition. Muscle fiber is strongly affected by any cooking temperatures, especially those over 130° F. Heat disrupts cell membranes, causing coagulation of previously discrete structures and allowing fluids to escape, liquefying the meat. By the time meat reaches 160° F, the temperature of well-done steak, most of the fluid that can be released is gone. Too much cooking after this point can turn muscle fibers tough and dry. But connective tissue requires higher temperatures to soften. Conversion of these tougher proteins to gelatin only begins at 140° F and is highest at water's boiling point, 212° F. So cooking meat to soften connective tissue can damage the muscle fibers, but cooking too little can leave meat tough to eat. Compromise thus is central to the art of cooking meat.

Out of this physical chemistry conundrum have emerged the varied ways of preparing meat. Tender sections of animals cook best with very different approaches from those for tougher cuts. A number of "dry methods"—roasting, baking, broiling, and frying—use high temperatures applied for relatively short duration to prepare well-marbled meats with tender muscle fibers and relatively little connective tissue. "Wet methods" such as boiling, stewing, braising, and steaming employed for longer intervals work best for tougher cuts where softening connective tissue is essential to make meat palatable.

Long before food science texts delineated meat's physical chemistry, the women who did most of the cooking were drawing on lore from their families, communities, and cookbooks to manipulate the meat cuts at their disposal. Use of particular wet or dry cooking methods, however, depended on available kitchen technologies.

Until the spread of iron stoves in the mid-nineteenth century, most women cooked over open fires on the household hearth. Pots could be suspended from a chimney bar that spanned the fire. Some cooking containers had short legs or were placed on "spiders" (metal holders with legs, also called trivets) so they could be situated over coals. Tin ovens placed adjacent to the fire baked food, while racks with skewers (sometimes with complicated gear arrangements) allowed meat to be roasted over the flames.

These pre-stove methods favored stewing and boiling meats, especially in poorer families who owned a limited repertoire of cooking equipment, and relying more heavily on cured meat. One colonial-era cookbook explained, "in general Boiling is the easiest way. To keep the Water really boiling all the Time, to have the Meat clean, and to know how long is required for doing the Joint or other Thing boiled, comprehends almost the whole Art and Mystery." Boiling and stewing were suited to heavily salted meats, especially pork. Hams, for example, would be soaked for more than a day prior to cooking to extract the salt from the flesh, and then tenderized through slow stewing. Not long before dinner, vegetables and other flavorings could be added.[2]

Roasting meat was associated with special occasions rather than conventional family dining, as the roast had to be freshly killed and closely attended during cooking. A smoky fire could ruin the meat's taste, as could the cook's failure to turn it at regular intervals; hence families employing servants to concentrate on these tasks or investing in automatic turning apparatuses were far more likely to have meat prepared in this way. Broiling steaks on a "gridiron"—a rack with iron bars on which the meat sat—had similar constraints and was also reserved for special occasions.

Frying meat entailed placing a pan on a trivet in front of the fireplace and pushing coals underneath. As a "coarse and greasy Kind of Cookery" it was associated with the "lower sort" in colonial society. Despite this association, frying remained popular because it allowed fatty salted meat such as bacon to be cooked quickly and easily.[3]

Cooking meat required considerable skill and knowledge for dishes at the lower economic levels as well as for elite repasts. Making a simple bone-based soup was no easy task over an open fire. One 1852 recipe published in Philadelphia for "winter soup" specified that its key ingredient, a shin of beef, needed to be sawed in several places to "cause the juice or essence to come out more freely." The shin was to be salted the day before boiling; then the cook was to place it, along with water, salt, peppercorns, and other flavorings, in the fire "as early as possible" in the morning. Once the water boiled, "the fire may be quickened," and other ingredients added around 9 A.M., principally cabbage and other vegetables. The soup cooked all day (requiring adept tending of the fire), and just before serving carefully transferred to a tureen so as not to include the "mass of shreds" of meat and bone as they looked "slovenly and disgusting" and also absorbed too much water. Leftover soup could be stored in "a tin or stone vessel" during very cold weather, but advice books warned women to avoid earthenware containers, as the lead used in the glaze "frequently communicates its poisons to liquids that are kept in them."[4]

This "simple," lower-class meal required considerable knowledge and work. Women had to obtain the proper ingredients by patronizing the appropriate butcher and vegetable stands; properly prepare and maintain the fire, ensuring there were sufficient supplies of wood and water (which needed to be carried in from outside the house); organize the meal preparation in conjunction with other household tasks performed during the same day; and properly preserve leftovers without recourse to refrigeration so that additional wholesome meals could be obtained from them. These women had to know how to navigate the fluid commercial markets at the same time that they had to be aware of the physical properties of meat so that it could be prepared in an appetizing and wholesome manner.

As iron stoves spread in the mid-nineteenth century, cooking changed to accommodate the new opportunities and limitations of this technology. The typical stove had four to six burners, a firebox in the front that usually took coal (some used wood), a large oven underneath the firebox, and in higher priced models, other accoutrements such as hot water reservoirs. By the 1860s there were sufficient low-priced models available so that working families could choose stoves over hearth cooking.

The physics of iron stoves made frying a more prevalent cooking method. Encouraging women to make proper use of the new stoves, an 1841 advice book commended the "round hole to which a griddle is fitted" that allowed the cook "to stand upright" while preparing food on top of the stove. Indeed, frying became so popular that *Harper's Monthly* regretted bringing the "unpleasant truth" to its readers that in many "places the national steak is—fried!"[5]

The ease of using surface burners and an oven elevated off the ground, and more efficient fuel utilization, pleased many women who shifted from hearth to stove cooking. But coal stoves were balky devices, especially for any use of the oven. Cooks could not see the progress of their foods inside the stove, and controlling temperatures entailed careful, exacting attention. The fire had to be monitored carefully, as opening the "drafts" to admit oxygen could lead coals to burn too fast, but limiting the drafts too much made it hard to reach good cooking temperatures. As coal was slow and difficult to light, enough had to be introduced to supply sufficient temperatures and cooking duration; on the other hand, putting in too much coal could burn the meat or even damage the stove by overheating the fire box.

Observation, essential for managing cooking over the open hearth, became far more difficult and dangerous with the coal stove. One cookbook advised that when broiling a beefsteak to first sear the meat by placing it "close to the hot coals and count to ten slowly; turn it and do the same," then after moving it back from the coals "turn it as often as you count ten" until it was done. One can imagine the cook wishing for the days of having a roast easily at hand, visibly cooking over an open fire. Traditional roasted dishes largely disappeared; the same cookbook noted that roasting meat before an exposed fire "is unquestionably the best method of cooking it." But as kitchens were no longer "equipped for roasting meat," baking "has come to be called roasting."[6]

Frying meats on a coal stove, while easier than over an open hearth, still required a great deal of attention and skill. Because the fat had to be the right temperature (and there were no mechanical indicators to ascertain such), housewives needed to learn the proper methods and visual signs. *The Century Cookbook* advised slowly heating the fat for one hour before use; otherwise, it would either be too cold to cook or have to be exposed to an open flame, "which is attended with great danger." Fat that was too hot and boiled over could catch on fire; in this eventuality *The Century Cookbook* advised using ashes to extinguish the blaze, and if the fire spread to the cook's clothes, to "roll on the floor until assistance comes." Assuming the housewife surmounted these dangers and brought the fat to the right temperature, items had to be lowered slowly into it to avoid splattering or making the temperature drop too far. Similarly, the food could not be left in too long, and had to be removed when it reached a nice "lemon color." Then the fat had to be brought back to the right temperature for the next batch, and so on. Cooked incorrectly, the food would be covered with grease, but contemporaries claimed that properly fried food "should contain no more grease than the boiled one does of water" and indeed that some meats were better fried. Similar to making a stew over an open hearth, use of coal stove frying to prepare meat was complex, requiring skill and sufficient knowledge of the cooking materials and technology. Aside from the problems of uneven temperatures, they were also awful to clean and presented terrible hazards to women who used them.[7]

As natural gas became more available at the end of the nineteenth century, homemakers were drawn to the stoves that used this new heating source to offer relief from the worst problems of coal. Some advantages are obvious: coal neither had to be physically loaded nor did the stoves require such exhausting cleaning. Less visible, however, were the dramatic changes in cooking methods these stoves generated. "The full heating power is developed from the moment of lighting a gas fire," *Scientific American* noted approvingly in 1894. "Scorching of food during cooking is completely provided against, since each burner can be turned down at any time, and the heat regulated to a nicety."[8] Heat was available with the twist of a knob, could vary between burners, and its level could vary; with such flexibility different cooking methods suddenly became possible.

Range manufacturers especially promoted the remarkable innovation of predictable cooking temperatures. "Prime roast beef is best prepared in a gas range, because the heat can be perfectly regulated,"

"COOKING WITH GAS

"SAVES TIME AND MONEY"

promised the Detroit Stove Works. "Wisely managed it promises perfectly cooked food," chimed in the makers of the Eclipse Range, as the "gas range is a coal range with a college education."[9] Economy, efficiency, effectiveness—these were persuasive themes to women tired of managing their coal stoves.

The ability to closely control the cooking heat, along with new measuring devices that allowed for precise recording of temperatures, changed cooking methods significantly. Cookbooks that assumed use of gas stoves specified precise temperatures and times in their recipes, quite unlike the nineteenth-century cookbooks guiding preparation of dishes in the open hearth or over a coal stove. While the spread of "scientific" home economics certainly influenced this approach, these new recipes would not have been possible without the stable temperatures gas cooking afforded. Temperature gauges on stoves, thermometers that could be inserted into meat, and inexpensive clocks made cooking a matter of timing and temperature adjustment rather than experiential judgment of how the meat was doing. One cookbook even warned of the inevitable failures experienced by women "who depend on their own judgement of oven heat" rather than depending on instruments that "are absolutely reliable, and will ensure uniformity" in cooking operations.[10]

The primary impact of gas stoves on meat cooking may have been an expanded array of "wet methods" to prepare the less expensive cuts, with attendant decline in stewing and boiling. Rather than cut up poorer cuts and put them in a pot over the fire or on top of a coal stove, thermostat-controlled ovens could be kept at lower temperatures to permit the slower cooking that allowed connective tissue to turn to gelatin while the muscle retained sufficient juice to be tasty. Braising recipes proliferated in guides like the *Good Housekeeping's Book of Good Meals* and the *Art of Cooking and Serving* (the latter published by Crisco manufacturer Procter and Gamble). Braising allowed relatively cheaper cuts like the rump or flank to be served as steak rather than stew meat. Cooks placed meat cuts on a small rack inside a covered baking dish, added water so that it stayed below the level of the meat, and cooked the food for three or four hours at 300° to 325° F, keeping the water steaming but not boiling. Some ranges also had a surface "simmering burner" designed so that a similar process of slow cooking could take place on top of the stove. Pot roasts relied on a similar approach, closely regulated ovens at relatively low temperatures to cook for a long time to soften the round. Such repasts certainly could be prepared on the open hearth or coal stove, but doing so was far more difficult because of these cooking technologies.[11]

Gas ranges were particularly suited to pan frying, perhaps accentuating the late-nineteenth-century taste for such methods—and encouraging butchers to imagine different ways of creating consumer meats. The simple ease of being able to turn on only one burner, rather than building a fire to heat the entire stove certainly made pulling out a frying pan to cook some meat a highly desirable cooking option. More properly called sautéing in nineteenth-century parlance, the lightly oiled pan was ideal for cooking chopped meat, reformed into new relatively thin shapes such as the meatball and newly popular hamburger. Frying also was a good way to cook tender meats like pork chops, the increasingly available spring chickens, or even thin Delmonico steaks.

It would take the postwar rise of the backyard grill to restore cooking over an open fire to American home food. With gas or electric ranges, broiling remained the principal dry heat method for choice steaks. Even here the impact of predictable temperatures was apparent, as precise time and heat levels replaced the "count to ten" approach of cookbooks oriented toward the coal stove. "Roasting"—known as baking to cooks raised on the open hearth—was for Sunday dinners or other special occasions.

The array of cooking options only tells us about available choices, not what was actually cooked. Any of these methods lent themselves to almost endless meal variations depending on spices, accompanying vegetables, sauces, and so forth. These cooking technologies are best understood as framing what was possible, encouraging certain cooking methods, and hence permitting consumers to get to the next step—determining what kind of meat they wanted for dinner.

How Much Meat?

Determining just how much Americans have eaten, and how that has changed over time, sounds simpler than it is. Yet while the exact numbers are hard to pin down, the general contours of Americans' meat eating habits are remarkably clear. Regardless of regional, ethnic, or racial variations, as incomes rose so did the demand for beef and poultry. Pork, in contrast, struggled to hold its own; from being the preeminent meat in the nineteenth century, it fell into third place by the beginning of the twenty-first century. And as meat provisioning and distribution became increasingly centralized on a national basis in the midtwentieth century, regional variations in consumption habits due to influences of climate and geography tended to fade even as cultural dimensions persisted. Convergence, in most respects, is the story of meat consumption trends in America up to the present day.

For early America the best studies have come from Sarah F. McMahon, who painstakingly reconstructed the meat possessions of families in Middlesex County, Massachusetts. According to her research, seventeenth-century farmers struggled to stretch their salted meat into the spring, but supplies stabilized in the early eighteenth century with the greater availability of salt and increase in domestic livestock. In the 1740s, 81 percent families in her study had preserved meat that lasted into the spring, with about half maintaining their supplies well into the summer. These proportions remained relatively the same into the mid-nineteenth century.[12]

Deducing the actual volume of meat consumption can only be inferred, as there were few true consumption studies until the twentieth century. One source is the meat allowance for widows (specified in wills), which rose from 120 pounds in the early 1700s to over 200 pounds by the early 1800s. While we do not know if these wills were implemented, such allowances indicate popular expectations of what constituted an abundant supply of meat; combined with other information we can surmise general practices. During the same period standard meat allocations for slaves in the South averaged around 150 pounds per capita. National estimates deduced from livestock-production statistics place per capita meat consumption in the 1830s at 178 pounds. These sources do give us some confidence in suggesting an average annual consumption of 150–200 pounds per person in the nineteenth century.[13]

Better census data (aided by the commercialization of meat production) combined with numerous microstudies of consumption habits in the twentieth century allow us to chart the contours of meat consumption over the last hundred years with greater precision. The British Board of Trade conducted the best early study of U.S. consumption patterns as part of a massive effort to prove the relative impoverishment of the British working class. A 1909 survey of more than 8,000 families in dozens of urban centers showed that per capita meat consumption (beef and veal, pork, lamb, sausage, and poultry) ranged from a low of 136.1 pounds for households earning under $1,000 annually to slightly over 200 pounds for families with incomes exceeding $2,000.

The 1909 study demonstrated that families increasingly favored beef and poultry as incomes rose, as would be the case throughout the twentieth century. Beef consumption rose 50 percent from the lower to highest income groups, while the amount of poultry tripled. African Americans were more likely to prefer pork than whites, but as income grew, they increased beef and poultry consumption far more than pork, in line with trends among white groups. Similarly, southern whites were more likely to eat pork than northerners were, but as income grew, their beef and poultry consumption jumped much more significantly than their pork consumption did.[14]

Immigrants' meat preferences did not differ dramatically from other white groups', even as they favored quite different spices and dishes. Slavic immigrants ate about 20 percent more pork and sausage than native-born whites, but beef still constituted their favorite meat by far. (Their total meat consumption actually was greater than other white lower- and middle-income groups.) To a large extent, the immigrant preference for beef reflected assimilation into America's meat-eating habits. A close study of working-class families in New York City's Greenwich Village area noted that while "foreigners bring their 'macaroni', 'bologna', or 'potato and tea' standards to America," the food budgets of these foreign-born families show that "they have become American-ized." All families studied consumed far more beef than pork, regardless of their economic status and ethnicity. One Italian family of peasant origin exhibited its ethnic roots by spending unusual amounts on macaroni, cheese, and vegetables. Nonetheless, beef expenditures were more than double that spent on distinctive "ethnic" foods and triple the amount spent for pork. The mother in the family made do, as poor New York housewives had since before the Civil War, by utilizing cheap cuts of beef to prepare stews and soups.[15]

Balancing the national trends favoring beef and poultry, regional traditions and practices particularly influenced pork consumption. Narrative descriptions of nineteenth-century consumption practices stressed the southern preference for pork, especially cured products such as bacon. An 1850 memoir of Georgia described "bacon, instead of bread," as the Southern "staff of life," something that could be found on the "Southern table three times a day." Pork also had strong rural associations, as it was easy for farming families to produce and preserve.[16]

These traditions wove their way into twentieth-century consumption patterns even as the growth of urban areas and the development of new preservation methods, principally refrigeration, altered the nation's provisioning geography. In the mid-1930s, beef provided only 12 to 14 percent of the meat consumed by whites and blacks in the rural South; pork consumption among these groups was four times that of beef. Similarly, white farm families in the North ate 50 percent more pork than beef; only one-fifth of their meat consumption was in the form of beef. Northern urban whites, in contrast, ate at least 50 percent more beef than pork. The rural-urban divide persisted in the West. While urban whites in the West chose beef for more than 50 percent of their meat supply—the highest percentage in the nation—beef comprised only 28 percent of the meat consumed by rural westerners. In the mid-1930s beef remained strongly associated with urban areas, while pork persisted as the dominant meat in farming regions, and dramatically so in the South.[17]

When Americans ate pork in the 1930s, region and race dramatically influenced the types they chose. Northern whites favored fresh pork over cured, and most of their cured pork came as bacon or ham. Southern whites and blacks were the only groups to continue eating salt pork, the cheapest form of cured pork, in considerable quantities. This variety constituted 17 percent of all meat that southern African Americans consumed, compared to less than 1 percent among northern whites. Differences even are clear among fresh pork choices; while whites preferred steaks, chops, and roasts priced at over twenty cents per pound, African Americans favored varieties simply labeled as "other" that cost

Graph 1.1. Annual U.S. Urban Meat Consumption, 1909–1965

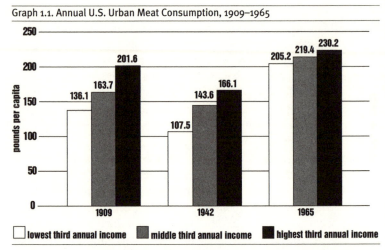

To standardize measures of purchasing power by income, I have used the cost of living index (CLA) to create equivalent income categories over time. I chose these gradients rather than food prices in order to establish the meat purchasing options of relatively similar income groups for 1909, 1942, and 1965. Taking 1909 as the base, the CLA was 50 percent higher in 1942 and three times higher in 1965. So the income gradients for the lower group were under 1,000 (1909), under 1,500 (1942), and under 3,000 (1965). The middle group was 1–2,000 (1909), 1,500–3,000 (1942), and 3,000–6,000 (1965). The highest group was over 2,000 (1909), over 3,000 (1942), and 6,000–15,000 (1965). The 1965 survey included higher income groups not included in the early studies, so I separated the over 15,000 category in 1965 to avoid skewing the comparison. This wealthier group, as might be expected, consumed considerably more meat, beef, and poultry than the 6,000–15,000 population. For similar reasons I have limited this table to urban groups, as national consumption levels are also affected by the increased urbanization of the U.S. population, not simply the increased access to meat of similar groups over time.

Sources: Great Britain Board of Trade, *Cost of Living in American Towns* (London: His Majesty's Stationery Office, 1911); U.S. Department of Agriculture, *Family Food Consumption in the United States, Spring 1942,* Miscellaneous Publication no. 550 (Washington, DC: U.S. Government Printing Office, 1942); U.S. Department of Agriculture Agricultural Research Service, *Food Consumption of Households in the United States, Spring 1965* (Washington, DC: U.S. Government Printing Office, 1966).

Graph 1.2. Annual U.S. Urban Beef Consumption, 1909–1965

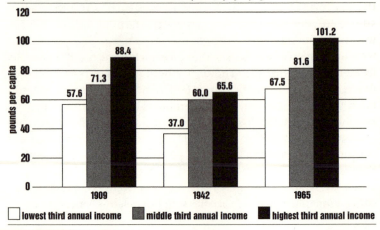

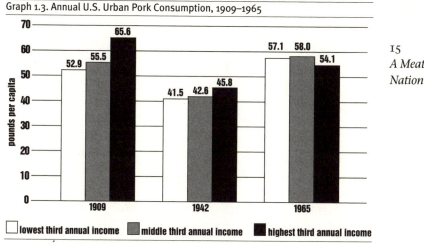

Graph 1.3. Annual U.S. Urban Pork Consumption, 1909–1965

twelve cents per pound, in all likelihood organs, feet, and the like.[18]

Meat consumption dropped to unusually low levels in the Great Depression. Mid-1930s household surveys document that employed whites in the North Atlantic region ate, on average, 122.4 pounds per year, southern blacks 110 pounds, and southern whites just 85 pounds. These studies only covered wage-earners in cities, excluding the huge numbers of urban unemployed and rural poor and thereby understating the sharp national decline. Estimates based on production data place the national average at 132 pounds per capita in 1935, much less than in 1909 but still concealing the actual drop in living standards.

Consumption rebounded in the 1940s, but it remained well below levels seen in the early twentieth century and was sharply distinguished by income. In 1942 Americans in the lowest third of income levels consumed just 107.5 pounds of meat per year, 20 percent less than in 1909. By contrast, Americans in the upper income group ate 50 percent more meat than the poorest third and consumed twice as much beef and poultry. Chicken remained a special meal associated with high incomes, eaten by only one-third of all families in 1948 but by more than 50 percent of households in the highest income bracket. At the other end of the spectrum, one-third of families with incomes under $2,000 ate salt pork, compared to only 10 percent in the higher income categories.[19]

Meat consumption began to climb dramatically in the 1950s after the end of the Korean War's rationing programs. By 1965 it had reached the highest level in American history with virtually all groups eating over 200 pounds per capita annually. Gains were especially pronounced at the lower socioeconomic levels; urban residents earning less than $3,000 annually still ate 205.2 pounds of meat per year. Income con-

Table 1.1. U.S. Meat Consumption, 1909, 1942, 1965

	1909	1942	1965
Beef	81.5	69.4	104.7
Pork	67.0	63.7	58.7
Poultry	14.7	20.7	40.9
Total	169.9	161.0	208.0

Source: U.S. Department of Commerce, *Historical Statistics of the United States* (Washington, DC: Government Printing Office, 1975), 1:329–31.
Note: Estimates based on production data.

tinued to matter, though, as the wealthiest strata of urban Americans consumed almost 50 pounds more per person than the lowest income group.[20]

Beef and poultry remained the meats of choice. Beef consumption almost doubled in quantity between lowest and highest income groups. Poultry (principally chicken) soared in popularity as lower prices allowed poorer Americans to catch up to the consumption levels of wealthier families. Per capita poultry consumption averaged 46.4 pounds in 1965, with less than 10 percent variation by income.

The 1965 data also shows convergence in meat consumption habits between city and country. Rural areas were now more closely aligned with urban preferences for beef. Nationally, farm families ate 50 percent more beef than pork, only slightly less than the proportion in cities. While country versus city differences faded statistically, the North-South divide persisted. Southerners in towns, farm regions, and rural nonfarm areas all ate more pork than the national averages in those categories. Pork had evolved from meat historically associated with rural areas to meat culturally tied to a particular region.

Within meat categories, however, socioeconomic status continued to matter. (The 1965 consumption surveys did not break down respondents by race or ethnicity, unlike earlier studies.) Wealthier groups ate steak twice as often as poorer families, and when they ate steak, they chose sirloins and porterhouse while the poor made due with cheaper (and tougher) round and chuck varieties. Differences also were apparent with pork, with upper income groups 50 percent more likely than lower-income families to eat fresh chops and cured hams. Sausage and bacon, however, once foods identified with the lower classes, grew in popularity at all income levels.

Contrasting the 1965 data with traditional food practices highlights the extraordinary growth in meat consumption following World War II, especially among lower-income families. Over two generations, Americans in the bottom third of income levels expanded their total meat consumption by 50 percent, with poultry increasing more than sixfold and better forms of pork and beef making their way into the home refrigerator. The postwar boom not only meant more cars and homes; it also could be measured by bacon in the morning, processed meats for lunch, and steaks, pork chops, and roast chicken for dinner. For the children who lived through the Depression's lean years and wartime austerity, obtaining ample supplies of meat, and varieties that were previously luxuries, was a daily indicator of postwar prosperity. For their children, who grew up in the meat-filled households of the 1960s, ample meat supplies were simply part of what it meant to have a prosperous America.

Meat's place at the center of the American diet has never been in doubt. When relatively marginal groups gained more buying power, their meat consumption grew dramatically, and they shifted their preferences to what most Americans already considered more desirable types of meat. The sheer cultural momentum is staggering; whether cooked over open fires or gas ranges, eaten in the homes of Slavic immigrants or African American sharecroppers, more meat and better meat were measures of the good life. Meat preferences emerged from the daily experiences of Americans seeking a good meal and were reproduced through intergenerational cooking lessons, daily meals, and celebratory feasts.

Yet Americans did not have an unlimited palate of meat options. Farmers may have produced their own meat, but the natural environment and preferences of meat processors limited what consumers could purchase. City dwellers may have had more variety, but they also had limited power over the purveyors who turned animals into food. Beef, pork, poultry, and processed foods came to the American table in forms its diners could not control—and often were quite unhappy with.

Neither did meat purveyors have a completely free hand in creating their products. The dratted animals took their own time to mature, demanded careful attention to fatten profitably, fit awkwardly into processing systems, and, once killed, could easily decompose such that the flesh could not be sold. Consumers posed problems too, as traditions governing meat preparation methods changed slowly, interfering with processing innovations that would make it easier to transform animals into meat. Out of this dance between producers, consumers, and nature would come the meat that Americans have eaten over our nation's history.

Chapter Two

Beef

In his early-twentieth-century story "A Piece of Steak," Jack London captured Americans' obsession with fresh beef. London's story concerns an aging boxer preparing for a last, desperate opportunity to support his family by fighting a rising young star. It opens with the boxer asking his wife if she had been able to obtain any meat for dinner. No, she explains, their credit at the butcher's was only enough for beef bones suitable for gravy. As the story proceeds, the boxer's nagging hunger haunts his efforts to win this all-important bout. Aware of his limits, the old boxer fights craftily, hoarding his strength and trying to tire the young man. He attacks when the opportunity seems right, using all his skill to quickly render a knockout. His blows push the young man on to the ropes, where he sways, one punch away from collapsing. But . . . the old boxer's energy fails—all because he didn't have that piece of steak. Unable to throw the finishing blow, the old boxer wilts before the young man's counterattack. The story ends with him sitting on a park bench, crying, unable to face his family and their impoverished life.

London's parable uses this family's inability to obtain steak as a barometer for social inequality. Poverty deprives this working man of the beef that will provide him with enough sustenance to support his family. His wife too is frustrated by her inability to secure adequate meat. In this story, beef functions not merely as a source of nutrition but as a symbolic expression of working-class aspirations for a better life.

Steak has been the symbolic pinnacle of American meat since the eighteenth century, and its preeminence has persisted virtually unchallenged for two centuries, just as the beef soup bones foisted on the boxer's family represented poverty. Meat supply and forms of meat have changed dramatically over American history, but beef has held a remarkably constant place in the nation's symbolic food universe.

Despite its popular appeal, beef's material properties inhibited consumption levels from equaling those of pork. Coming as it does from large animals, beef was an inconvenient meat for rural families as so much needed to be preserved. Its physical composition—principally longer muscles and connective tissue and relatively lean composition—

made cured beef relatively unpalatable. The necessity of preserving large amounts to make efficient use of the cattle's meat warred with Americans' persistent taste for fresh beef.

These intrinsic tensions had a defining influence on patterns of beef production, distribution, and consumption. Beef eating in America was more prevalent in towns and cities that could support steady demand for the fresh cuts of this meat. In order to find customers for the entire animal, urban purveyors refined a wide spread of products that appealed to different markets. The best roasts and steaks were America's most expensive meats and barrel-cured beef the cheapest, with many gradations in between. Encountering meat in this way—in public markets and butcher shops—urban consumers and shoppers were daily imbued with the intricate differences in cost and status associated with different beef cuts. Then (as now) no other meat had beef's variety of forms. Rural Americans were simply far less likely to have access to beef at all, except for special feasts when sufficient numbers gathered to justify slaughtering an 800-pound animal. For similar reasons, beef consumption was highly shaped by climate and more prevalent in the North and West, where prolonged cold winters and ample cattle supplies made it easier to keep fresh meat palatable.

Beef purveyors adapted their methods to these consumption practices. Until the adoption of refrigeration in production and distribution, cattle were slaughtered close to their place of sale and in extremely small numbers. Once cooling technology permitted large packinghouse operations to emerge in the 1880s, packers eschewed development of cutting operations similar to those in the pork industry which prepared meat for curing. With the focus still on delivering fresh meat to consumers, beef slaughterhouses limited operations to killing, splitting, and chilling, and then shipping beef in 200–250 pound "quarters" to local butchers—first in public markets, later in private shops—who then fabricated cuts in forms that met local preferences.

Centralizing beef slaughtering in the American Midwest, with attendant national distribution systems, made beef America's principal meat by the twentieth century. As the nation became more urban, the refrigerated trains (and later trucks) conveying beef quarters to wholesale and retail outlets allowed communities without slaughtering facilities to share in the national appetite for fresh beef. Production remained constrained by consumption preferences, however, as the challenge of turning these large animals into attractive consumer cuts mandated a highly skilled, manual production process. The leverage thus accorded to slaughterhouse workers and retail butchers ultimately permitted unions to take a strong hold in the red meat industry in the mid-twentieth century.

Local Beef

Country slaughtering of cattle for local consumption was a simple process. All farmers needed was a means to secure the animal for the killing blow (a sturdy stump or post would do), a place to hang the carcass while it was "opened" and the intestines removed (a tree could suffice), and a way to dispose of the animal's innards and other inedible parts (often a local stream). The problem then came with what to do with the meat.

Despite beef's appetizing appeal when eaten fresh, it was cumbersome to cure and generally did not taste as good as salted pork because of its tougher fibrous nature. In rural areas of the North and Midwest, itinerant butchers established set routes and sold beef to dispersed homesteads, but limited transportation and the absence of home refrigeration were impediments to developing a large clientele. (Women purchasers also were understandably dubious as to the wholesomeness of fresh meat kept under canvas for several days in warm temperatures!) Sam Hilliard has carefully documented a large surplus of beef cattle in the South, indicating that even though southerners had plenty of opportunity to eat beef, they generally preferred pork. In urban areas such as New York City, however, fresh beef was clearly the most desirable meat. Butcher Thomas De Voe decried the preferences of New York shoppers who "would rather pay their last dollar for half as much meat in an expensive steak or chop" than purchase inexpensive cuts. It was these urban areas with public markets and eating establishments that set the pattern for beef consumption in America.[1]

Urban consumers' demand for beef stimulated cattle and livestock raising in rural areas capable of supplying this market. New England farmers played the livestock markets astutely, investing in equipment and land for these purposes well before the American Revolution.

As New York became the nation's leading city in the early nineteenth century, it also assumed central place in the East Coast rural-urban livestock nexus. New Jersey livestock producers assembled cattle at convenient shore points and transported them to Manhattan by ferry. Similarly, "upstate" farmers in Westchester and Dutchess Counties relied on drovers to assemble herds and drive them to the Hudson River, whence they were brought down to the city by sloops. Cattle also came from Connecticut and Massachusetts, if New York prices exceeded those in Boston, and, increasingly after 1820, from the West, principally Ohio. Drovers brought the cattle by sailboat or ferry to the city's central stockyards on the Lower East Side, traditionally located behind the Bulls Head Tavern. Butchers carefully selected and purchased animals from the drovers and transported the animals to small slaughterhouses located throughout the city. Once the animals be-

Catharine Street Market, lower Manhattan, 1850

D. T. Valentine, *Manual of the Corporation of the City of New York for 1857* (1857). Courtesy Hagley Museum and Library

came property of the butchers, these men took control of the urban meat trade.

New York's antebellum urban butchers retained a dual identity as craftsmen and small shopkeepers. A limited number received licenses from the municipal government to sell meat in the only legal locations—public markets scattered throughout the city. Butchers paid a small fee to kill animals themselves in small slaughterhouses, or left that task to their apprentices and journeymen. As there was no refrigeration available, summer killings took place at night so that the meat would not spoil before coming to the market. The fresh meat was brought to the public markets around dawn and prepared to order for customers. Often the butcher's wife handled the transactions while the butcher did the cutting, with the apprentice then conveying the meat to the customer's residence. The stalls were simple affairs, with a wooden table upon which to cut meat and a few hooks on upright posts to hang the joints and other large pieces.

European visitors were astounded by the vast array of products available in New York's markets. A British visitor granted that while an Englishman might miss the "prime fed beef of Smithfield . . . the abundance of good things, and their comparative cheapness, satisfy him that he is in the land of plenty." Limitation of meat sales to public markets ensured that the licensed butchers remained in control of municipal meat supplies through the first half of the nineteenth century.[2]

Most nineteenth-century city dwellers could obtain beef in some form. Travelers' accounts document the opulent beef consumption habits of the upper class, whose members might eat several types of roasts at one meal. But it was the quantity and type of cuts that distinguished elite eating habits from those of common people, not beef consumption as such. An 1851 food budget from a working-class family indicated that with the cost of meat averaging ten cents a pound, annual per capita consumption of "butcher's meat" by family members was 146 pounds. This budget probably reflected conventional usage of cheaper cuts for stewing and boiling, employment of bones to make soup stock, and an occasional roast or steak purchased for special occasions. While not distinguishing between beef and pork, the reference to "butcher's meat" attests to the availability of fresh meat, including beef, to urban residents of modest means.[3]

Class nonetheless heavily influenced the types of fresh beef eaten by nineteenth-century Americans. At the pinnacle of the beef hierarchy were roasts such as the "baron of beef," a cut of English origin served "as the crowning dish for the Christmas dinner" of the reigning British monarch. It was formed by separating the loin from the rear hindquarters of a "beeve" that had not been split down the middle; this formed a large, square, and boneless piece often weighing more than one hundred pounds. Americans, even in elite circles, found this cut extremely unwieldy to prepare and instead were fonder of cuts that separated the "baron of beef" into smaller sections to form sirloin or rump roasts, often weighing more than twenty pounds. Almost as popular were roasts cut from the first nine ribs of the forequarters, "considered by many epicures to be the finest and best-flavored parts of the animal." As these roasts cost between 15 and 20 cents a pound, they were generally the province of the upper class.[4]

Steaks cut from the loin also were popularly served at New York City's taverns and eating establishments and broiled over open fires rather than roasted. New York butcher Thomas De Voe attributed the term "porterhouse steak" to the fortuitous decision of a tavern (or "porter house") keeper to serve a famished boat pilot a steak cut from large sirloin originally intended for roasting. Although this cut was chosen in desperation, because he had run out of his usual stock, customer reaction was so positive that soon butchers throughout New York's markets were cutting meat in this way for tavern owners. While De Voe's tale cannot be verified, it reflects how beef eating was an important part of masculine public life and accessible to working men who perhaps could not have afforded roasts but still could buy smaller cuts of premium beef. It also indicates the close relationship between consumer demand and provisioning practices at a very early stage of the American meat industry. In this vernacular tradition may lie the origins of London's parable about a "piece of steak."

Most urban residents, however, could not afford the pricey broiled

Thomas De Voe,
c. 1865

Thomas De Voe, *The
Market Assistant*
(1866). Courtesy
Hagley Museum and
Library

steaks and roasts for home consumption. Other, less desirable, forms of
beef had to suffice for the stews and soups that constituted more conventional cuisine. Tougher cuts, such as the flank, rounds (both from
the hindquarters), brisket, and plate (the latter two from the forequar-

ters) generally served for stews, as longer cooking times in water softened them sufficiently. Bony meat, such as the neck, shoulder, and thigh, was "excellent for a sweet, strengthening soup." Beef livers and kidneys also were used for stews and soups, and De Voe particularly endorsed the heart as making a "good, wholesome and delicious meal" even though it was "one of the cheapest [parts] in the animal" and could go for as low as four cents a pound. Poor residents even could obtain beef shins, though they were "fit for nothing but stock for soup."[5]

The brisket and plate—both lean cuts—were the typical sources of cured beef. These were low-grade products in comparison to most forms of cured pork. While salting pork rendered it "perhaps as good as any can be," an 1805 farmers' almanac observed (reflecting popular taste) that "beef is greatly injured, and rendered unwholesome, by a severe salting." Most cured beef was "corned" through a wet-cure method similar to barrel pork, where it was placed in a salty brine solution and kept in the liquid until ready for eating. Wet cured beef could last for close to one year, similar to pork. Drying or "jerking" beef kept it wholesome for longer periods. Although nutritious, this hard, tasteless product was one of the least desirable forms of meat.[6]

A tongue-in-cheek "exposé" on New York City boardinghouses drew on popular attitudes toward beef to demarcate among types of these residences. Establishments catering to the poorer sort obtained tough "boarder's beef" for its meals, meat allegedly selected from an animal "during whose long and useful career plentiful exercise has compensated for inadequate nourishment." The "cheap hotel boarding-house" served these "coarse meats" for its workingmen clientele as they valued "quantity rather than quality" when it came time for dinner. Indeed, the quality of beef distinguished rough establishments from the genteel. In one of the latter, the "proprietress" was one of the most "popular of landladies" as her "temper and beef were beyond all praise."[7]

New York's skyrocketing population (exceeding one million in 1850) created accelerating demand for meat supplies. Such rapid growth also created intractable problems. The dispersed, ubiquitous meat business practiced in close proximity to the public markets and residential areas became a public nuisance as production expanded. A European visitor in 1830 was appalled that the city's slaughterhouses were "scattered over many populous districts of the city." He observed that these establishments saturated the air with "the most noxious effluvia" and deplored the visibility of "the necessary but disgusting business of blood, which is carried on in buildings exposed to the public thoroughfares." Thomas De Voe, who lived at the intersection of Spring and Mulberry Streets in 1822, recalled that there were twenty-five or thirty slaughterhouses within 200 yards of his home. Though a booster of the meat business, he admitted that "when there was a southerly breeze" the slaughterhouses located along New York's back streets "would acquaint you of their close proximity."[8]

By the 1840s the city's slaughterhouses (numbering over 200) were under attack from an urban public health movement. The "absorbent or defective floors" of these small operations, complained the New York Board of Health, allowed blood and other animal refuse to pool or sink into the soil and give off "offensive odors." Further, parts of the animal "not destined for human consumption" were "separately disposed of" by "petty tradesmen" who, by carting these animal parts to their marginal establishments, conveyed the "sickening stench" of the meat business throughout the city. After several decades of protests the Health Department finally forced slaughterhouses to relocate to the city's east and west waterfronts north of 40th Street.[9]

Public health concerns intersected with growing discontent among women shoppers unhappy with the difficulties of obtaining meat. For those without home refrigeration, shopping was a daily chore, and as the city expanded new neighborhoods were increasingly distant from the dozen public markets where meat could be sold. A 1797 petition asked for two local men to have the opportunity to sell meat because of "the distance your petitioners live from the now established public markets." Forty years later a similar appeal led to the establishment of a Harlem public market on a site previously "indifferently supplied" by "meat-wagons" that visited on an occasional basis. While the women who did most of the shopping could not express their displeasure at the public market system by voting, they exercised considerable political influence by patronizing wagons peddling meat that plied regular routes through their neighborhoods, albeit breaking the law to do so.[10]

The butchers fought valiantly but unsuccessfully against these perceived assaults on their status. Beginning in the late 1830s they tried to secure city council measures to close the many unsanctioned meat shops "now established in almost every part of the city" that were operating "with perfect impunity." They deflected claims of a monopoly by defending the public markets as an ideal location to promote true competition, and also to ensure the quality of meat for the public. Rather than open up private shops that "are calculated to become the vehicles of vending bad and unwholesome provisions," The butchers urged the city to create smaller and more widely dispersed public markets that would alleviate the shopping burden.[11]

The butchers' efforts were to no avail. In 1843 the city council legalized meat sales in private stores. By 1850 there were 531 butcher shops scattered through the city in addition to public market stalls. That year an observer, obviously struck by the dramatic change in meat retailing in less than a decade, commented, "it has become fashionable to have a meat shop on almost every corner." Following the Civil War, with

killing operations pushed north of 40th Street, the butchers who operated stalls in the remaining public markets worked in "a great wholesale meat emporium" dispensing meat to butcher shop operators rather
than individuals. Butchers fragmented as a social group into retail
butchers operating small stores, packinghouse workers laboring in the
"slaughterhouses near the wharves, uptown," and the owners of wholesaling establishments.[12]

Nineteenth-century urban geography established the basic hierarchy and structure of American beef products. In the give and take of
face-to-face transactions, butchers learned the value of cuts and how to
refine them for local taste. Steaks for roasting or grilling were the most
desirable pieces, delineators of status for the wealthy and markers of
special occasions for those of more modest means. Seeking to vend less
desirable products, butchers sought to satisfy requirements of traditional recipes that softened the tough cuts or extracted valuable proteins from bony parts of the animal. The specificity of local production
and taste promoted considerable variation in cutting methods from
city to city, even as the beef hierarchy remained remarkably stable and
consistent. The separation of slaughtering from retailing, however,
presented great opportunities to transform the means for supplying
larger numbers of Americans eager to obtain a piece of steak.

National Beef

The beef business began to change as a result of American territorial
expansion into the Southwest. Acquisition of Texas and the adjacent
grasslands through the 1848 Mexican-American War encouraged expansion of commercial cattle raising, primarily through taming the
wild longhorn cattle. Following the Civil War, extermination of the
buffalo herds and expulsion of the plains Indians from this region
opened up huge areas for this new livestock industry. The ramifications
of these developments would reverberate from Chicago's stockyards to
New York City's meat markets.

The initial phase of nationalizing beef was an explosive growth of
the live cattle shipping business. The first Texas steers arrived in Chicago in the late 1850s, and progressively larger numbers came to the
stockyards for the next twenty years. The cattle generally were not killed
in Chicago. Driven to interior railheads like Abilene, Kansas, they were
live-shipped "on the hoof" by rail through the Chicago rail hubs to eastern slaughterhouses. Between 1864 and 1880, companies such as the
Pennsylvania Railroad built a network of stockyards at which the cattle
could be fed, watered, and sold to local slaughterhouses. Its livestock
shipments quadrupled between 1859 and 1863 and continued to rise
steadily. In 1866 Pittsburgh's East Liberty stockyards dispatched 14,284

cars headed east; a decade later departures of cattle cars reached 64,773, almost 200 per day.[13]

The live cattle market was a boost not only to Chicago's stockyards but also to the modest New York City slaughterhouses. Despite the
Board of Health's objections, dozens of small slaughterhouses remained in operation well after the Civil War, clustered in close proximity to the docks where ferries unloaded their cattle. Hopes that the Pennsylvania Railroad's slaughterhouses across the river in New Jersey would put the New York operations out of business were dashed by consumer preference for beef killed as close to sale as possible. In 1880 New York was still the nation's leading beef-producing center. Until the natural deterioration of meat could be halted by means other than curing, the beef industry remained limited by climate.

Modest growth in midwestern slaughtering accompanied the explosion of the live shipping business. But the Chicago plants killing Texas cattle differed only marginally from New York's slaughterhouses of the same decade. Neil Carbrey (a butcher in the 1870s) recalled that the typical Chicago beef plants of that era had a gang of around fifteen men who slaughtered but ten animals per hour. The cattle were killed by a blow to the head while secured to a bull ring, and then "pumped" by moving the rear legs to push blood (not very effectively) out of the animal's circulatory system. While the carcass rested on the ground, "floorsmen" wielding long knives separated the animal's hide from its stomach and sides, aided by workers who used a hand winch to elevate the animal and turn it over. "Hoist men" then lifted the cattle off the ground for the finishers to make the last few cuts in the flesh holding the hide to the carcass, and then pulled it higher into the air so that "splitters" could sever the carcass in two and "headers" remove the skull.

As in antebellum New York City, these crude prerefrigeration beef slaughterhouses were vile establishments. "It was customary in those days to wipe your boots with the same cloth we used to wipe or wash the beef," recalled Carbrey. In lieu of dressing rooms, workers were assigned cattle sheds to change clothes. As these plants were little more than wood shacks, workers shivered through the winter and wore as little as possible during Chicago's steamy summer. The offal and excrement that littered the plants did not trouble the workers, who would clean off by jumping into the Chicago River—where the plants' refuse also went. Without refrigeration the cattle were sent to the public market the day after they were killed; in the event business was slow, they were held over for the "Libby's man," the agent of the leading beef canner, who would buy them at a discount.[14]

All this changed soon after Gustavus Swift moved to Chicago in 1875.

Swift was a true innovator and the first packer to successfully apply refrigeration to large-scale beef production and distribution. By increasing the amount of time that meat could be kept fresh, refrigeration altered the geographic options for meat producers. This innovation fell far short of "annihilating space" as William Cronin suggests in *Nature's Metropolis,* as the meat still remained highly vulnerable to deterioration if not handled properly. Nonetheless, loosing the constraints of climate had dramatic consequences. First, processing took place on a far greater scale. Packers now could take more time to sell their meat, as the natural process of decay could be held at bay, allowing a greater volume of production. Second, using ice to prevent spoilage made it possible for packers to separate slaughter from sale. While meat's natural character previously had bounded firms' strategies within rigid spatial and temporal constraints, refrigeration opened up new arenas of competition, as well as opportunities for technical and organizational innovation.

It took an experienced meat man like Swift to broach these barriers. A product of the local East Coast meat business, he was a skilled all-around butcher in the tradition of Thomas De Voe. Swift had been active in the eastern Massachusetts fresh meat business for twenty years prior to moving to Chicago. He began his career as an itinerant meat wagon driver who sold cut beef to farmers and working-class families along a regular route. Like De Voe and other antebellum butchers, the young Swift routinely inspected and purchased his own livestock, killed and dressed the animals, and then sold the meat to regular customers. He was one of many small dealers who frequented Brighton stockyards that supplied many Boston-area butchers with livestock. Swift would buy a cow (and sometimes a pig or two) at Brighton on Thursday when it was thronged with hundreds of livestock dealers and meat purveyors. After slaughtering and butchering the animal, Swift used Friday and the weekend to peddle the meat from his cart, returning again to Brighton on Thursday to repeat the cycle.

Immediately after the Civil War, Swift expanded his operations by retiring his wagon and opening a meat market in Clinton, Massachusetts. He continued to frequent the Brighton market, however, buying, slaughtering, and butchering the animals himself. Swift's considerable cattle-buying skills brought him to the attention of two large meat dealers in the Boston area, one a slaughterhouse operator and another a shipper of live cattle to England. Looking for new opportunities, Swift accepted their offer to begin a new trade as a cattle buyer and turned the retail market over to a brother.

This career decision would lead Swift to the Chicago stockyards, and to a revolution in the meat business. In his new job Swift began to follow the flow of live cattle to their source, seeking the best animals for the best price. As western beef was now supplementing the local supplies of cattle at the Brighton market, Swift began to investigate the new distribution chain established by the railroad companies. He followed this stream backward, first to railroad stockyards in Albany and then to a far larger operation in Buffalo, and finally to the Union Stockyards in Chicago, where he arrived for the first time in 1875.

In Chicago, Swift saw firsthand the locational and transportation advantages that had made it a dynamic meatpacking center. The livestock trade was booming, fed partly by the influx of Texas cattle shipped by rail from Abilene and Kansas City. He also learned about the use of refrigeration in packing operations. By 1875 ice-cooled storage facilities permitted hog processing to take place year round, and entrepreneurs like G. H. Hammond were trying, so far unsuccessfully, to develop chilled railroad cars to ship fresh beef to the East.

Swift started by sending cattle "on the hoof" from Chicago to his associates in the Boston area but immediately began pondering an obvious question: wouldn't it be more economical to ship dead meat than live steers? It was expensive to feed cattle on the train trip; many died, and most lost weight. Moreover, only 60 percent of a cow could be used for meat. Clearly it was more economical to ship meat after it was butchered—but how to do so with this perishable product?

The critical obstacle was developing a railroad car that could keep meat fresh for several days. This was simultaneously a technological and organizational problem. A car had to be devised so that the meat stayed cool, but did not come in direct contact with ice. And a means of replenishing the cars along the way had to be devised, as the journey east took several days and could not be accomplished with a single load of ice. It took Swift several years to overcome these obstacles. By 1878 he had secured patents on a car that could keep meat cold even in the summer, a contract to have these cars built on credit, and a railroad willing to carry them east. To maintain these cars during their voyage, Swift built ice stations along the train route and obtained ice harvesting contracts on the Great Lakes.

By 1880 Swift's dressed beef was reaching the East Coast, with its initial breakthrough coming in the Boston area through his brother and other business associates. Within just five years the tonnage shipments of this new product out of Chicago had pulled even with shipments of live cattle to eastern markets. Relatively inexpensive "western" dressed beef was widely available in New York, Philadelphia, Baltimore, and of course Boston, ushering in a new "era of cheap beef" in the celebratory words of *Harper's Weekly*. At the same time eastern stockyards went into a decline from which they never recovered. Sales at Brighton stockyards in Boston, where Swift had bought cattle for twenty years, fell by 50 percent in the same five years.[15]

A problem remained, however: the refrigerated car only provided transportation—how to get the meat to consumers? In East Coast cities retail butchers dominated meat distribution and were accustomed to selecting their meat from local slaughterhouses if they did not kill the animals themselves. They resisted the entry of "chilled beef" into eastern consumer markets, quietly supported by the railroad companies who had made a great deal of money conveying live cattle and had substantial investments in transshipment stockyards.

To obtain outlets for his product, Swift created a proprietary distribution system of branch houses and railroad car routes to complement his extensive shipping operations. Swift's meat went directly from the packinghouse into his wholesaling network, which exceeded 200 outlets by 1900. Local butchers could purchase their supply of meat from a Swift branch house in their town, or from a refrigerated car that made regular stops along rail lines. Swift defused local opposition by investing in local meat distribution companies and making allies out of former opponents. And the low prices for "chilled beef" quickly overcame questions consumers may have had about western meat. Just five years after *Harper's Weekly* promised an era of cheap beef, prices had declined almost 20 percent; over the next five years prices dropped even further to $4.75 per hundredweight, 30 percent below 1882 levels and on par with beef prices in the 1840s.[16]

Other midwestern-based firms emulated Swift's strategy and invested heavily in integrated packinghouse expansion, refrigerated cars, and a distribution network. Their packinghouses located in urban centers able to draw on rural livestock supplies grew into sprawling multibuilding operations as firms expanded production and added new product lines. Swift's Chicago complex had grown to twenty-seven buildings by the time of Chicago's 1893 Columbian Exhibition, and Armour installed an elevated electric railroad to transport meat between its many separate structures. Geographic expansion accompanied growth of Chicago operations as firms opened plants in midwestern centers such as Kansas City and Omaha and established international operations in beef-producing regions such as Argentina. By World War I their network of branch houses and refrigerated car routes reached into 25,000 communities in America, and international distribution operations acquainted customers throughout Europe and Latin America with the products of America's leading meat companies. Offices employing hundreds of clerks tracked these operations, as bureaucratization of company operations paralleled continued family control by the children of the firms' founders.

The power of these companies over the production and distribution of meat allowed them to dominate the American market and displace local meatpacking concerns. Unless companies had similar resources, they were forced out of the fresh meat business or relegated to local spe-

Swift & Company Chicago complex, 1893

Warshaw Collection of Business Americana—Meat. Courtesy Archives Center, National Museum of American History, Behring Center, Smithsonian Institution

cialty markets such as the kosher beef trade for New York City's large Jewish population. By the time of Swift's death in 1903, six firms controlled 90 percent of the inspected cattle slaughter in the United States. Packing companies moved into the distribution of soft drinks and other grocery products that were well suited for distribution through their branch house network. By the early twentieth century Swift and other large firms were poised to become purveyors of the nation's food, not simply its meat.

The political troubles experienced by meatpacking firms in the early twentieth century stemmed directly from their enormous market power. A Federal Trade Commission investigation during World War I demonstrated conclusively that leading firms had conspired for over thirty years to fix prices and allocate market share. To avoid federal sanctions, the leading firms agreed to a consent decree and the federal Packers and Stockyards Act regulating their activities. The companies terminated distribution of nonmeat items through their branch house network, precluding these firms from moving aggressively into general retail food distribution.

Nationalizing beef production through the rise of the Midwest's meat industry made beef more widely available for Americans. With the branch houses and refrigerated car routes extending throughout the nation, consumers no longer had to rely on local livestock sources to obtain beef. In 1909 total beef production (including veal) topped pork products for the first time in American history with more than 4 billion pounds of fresh beef leaving the nation's slaughterhouses, a 44 percent increase in ten years. America had become a beef-eating nation.[17]

The beef hierarchy articulated by Thomas De Voe persisted largely unchanged into the twentieth century. Because beef was eaten fresh, minute differences in texture and fat content distinguished between varieties. "There is as much difference between beefsteaks as between faces," counseled an 1899 cookbook. A comprehensive food retailer's encyclopedia published in 1910 advised that the best beef "should be of a fine, smooth texture and bright fresh red color intermixed with fine streaks of white fat." Cuts from the loin and rib were the most desirable forms of beef because of their mixture of fat and lean. The general public's belief "that there are only three or four pieces of an entire beef that are fit for the table" frustrated twentieth-century butchers as much as it had Thomas De Voe in the 1850s. It would be a "revolution in ideas" thundered the author, "when heads of families realize that there are many cuts of beef equally as nutritious as the sirloin, porterhouse steaks, and standing rib roast, which can, with very little extra trouble, be served in forms just as palatable and inviting."[18]

In practice, beef-eating Americans remained willing to eat meat from all parts of the cattle, seeing fresh beef as itself containing many varieties and taste options. "A majority of the diners call for beef dishes," observed an experienced hotel chef in 1936. "You may eat beef three hundred and sixty five days in the year and not tire of it. You cannot say that for any other kind of meat." Yet steaks and roasts remained intractably perched on the top of the status hierarchy.[19]

The home economics movement was one source of advice encouraging homemakers to look beyond steaks and roasts as desirable beef dishes. For soups and stews a 1927 guide prepared by two leading home economists advised using the neck ("Juicy and well flavored"), hind, and fore shanks. The "fat is sweet" from the plate and navel, making them good for "forming a part of boiling meat." Ribs unquestionably made the best roasts and loins the best steaks, but the chuck could be used for both too, as it was a "Good quality meat" despite being "low in fat." And housewives looking for a bargain were advised to consider the rump for roasts, steaks, and stews because it was a "solid, juicy piece of meat of good quality." Despite this sage advice, the authors inadvertently indicated their acceptance of the traditional beef hierarchy by responding to their own rhetorical question "why we like meat" with the answer, "After a meal with a juicy steak or roast we feel well fed."[20]

Beef producers, the packinghouse firms, had an attenuated relationship with consumers and little ability to directly influence their preferences. Consumers bought their beef at butcher shops and were unaware who had produced it, unlike pork whose processed forms were clearly branded with their manufacturer. Butchers either prepared beef to order, or, if displayed already cut, wrapped pieces in brown paper at the request of the customers. A butcher's shop would obtain its beef from packing firms' branch houses, either by visiting personally or through deliveries made to their store. Meat firms learned of consumer preferences through retailers, who in turn were influenced by their interactions with customers.

A complicated balance of trust and tension characterized the point-of-sale relationship between retail butcher and shoppers. Studies conducted in the 1920s and 1930s agreed that customers bought meat from stores that were very near them, generally within 1,000 feet or two blocks, and patronized the same store for years at a time. These establishments were small; of the 44,000 butcher shops identified in the 1929 census, over half had sales of less than $20,000 annually. Manhattan's 2,000 meat markets averaged but $35,000, and most consisted of an owner and one employee. Nationally the 105,000 stores that sold meat along with other food products were slightly larger, but a majority still had annual sales of under $30,000. Meat departments located in chain stores accounted for at most 15 percent of all retail meat sales in 1929. Overwhelmingly shoppers obtained meat in small stores near their homes where they had established relationships with the proprietor and his employee.[21]

The women who bought meat were discerning shoppers. In consumer surveys shoppers consistently ranked meat quality as the principal reason for patronizing a particular store, though this was equaled by price concerns among poorer customers. A 1930 study of Pittsburgh meat shops ascertained that 60 percent of shoppers relied on a combination of the meat's appearance and odor to make purchasing decisions. Looking for meat that was red, fat that was white, and cuts that smelled fresh was, for most shoppers, the first step in making a beef purchase. A national study conducted in the mid-1920s showed that these qualities were especially important in steak and roast purchases, as they were measures of these cuts' "palatability."[22]

These preferences posed a problem for the local butcher shops. As beef came in cattle quarters, the character of consumer demand forced

them to accentuate price differences among beef varieties in order to
not be left with parts of the animal. Elite beef such as porterhouse steak
sold for 55 cents a pound in Pittsburgh around 1930 and a rib roast went
for 40 cents, but flank steak sold for 33 cents and boiling beef 26 cents
a pound. Confident of steak and other high-end sales, chain meat markets tried to encourage customers to go for other varieties such as chuck
roasts, which were advertised more than any other meat item in the late
1920s. Such pricing strategy further reinforced steaks' and roasts' place
as the most elite and desirable meats.[23]

Shoppers' preferences, especially insistence that purchases have the
proper physical attributes of color and smell, were powerful influences
on the practices of meat retailers and producers. Butcher shops had
to keep meat wholesome and, perhaps even more important, looking
"fresh." Meat retailers used ice-chilled coolers for this purpose following the Civil War, with many shifting to mechanical refrigeration in
the late 1910s and 1920s as equipment suppliers began marketing units
for small store use. A "Butcher Special" refrigeration system sold by the
Baker Ice Machine Company promised to automatically maintain constant internal temperatures and thus deliver "vastly better results than
the sloppy, irregular, unsanitary ice-cooling method."[24]

Display case manufacturers especially boosted the sales appeal of
meat kept under mechanical refrigeration. "If customers see your perishable products in their full and natural colors, neatly arranged in display windows and showcases," asserted a 1920 Fairbanks, Morse catalog, "they are naturally impressed" with the store's commitment to
"quality and freshness." The value of glass-enclosed coolers "so that

merchandise can be temptingly displayed to customers without spoilage in the very hottest weathers" was a constant theme among mechanical refrigeration vendors. "Your customers are the same people that buy from department stores," lectured a York Ice Machinery catalog.

"They like to window shop, and are influenced by what they see." In addition to the "sale-stimulation value" of the visual "array of choice meats," mechanical refrigeration prevented the odors produced by the "wet, musty, and unsanitary" ice refrigeration systems. These manufacturers and their equipment customers would have endorsed the results of the marketing survey that appearance and smell were the most important factors affecting meat sales.[25]

A 1929 meat retailing study indicated that stores were responding to the entreaties of mechanical refrigeration suppliers. Larger, prosperous meat stores "have modern equipment with extensive refrigerated show cases, many meats already cut in trays, and prices conspicuously displayed in order to facilitate the rapid service of a large number of customers." Their modern equipment included "sectional boxes, refrigerated show counters, window refrigeration, and grinders and slicers." Meat dealers with around $70,000 in annual sales had an average of 23 feet of refrigerated show counters, having added five to seven feet in the prior two years.[26]

The color, texture, and appearance of beef in the butcher shops, however, could only be as good as the meat that left the packinghouse door. Pressure from the consumers transmitted via the retailers meant that packing firms had to establish a price incentive system so that farmers would deliver animals with the desirable fat-laced beef that fetched higher prices. Firms developed elaborate private grading systems in the early twentieth century to distinguish the quality of animals used for beef. The American Meat Institute of Trade identified nine different categories, ranging from the eating grades of prime (first), choice (second), and good (third) to the cutter (eighth) and canner (ninth), which fetched far lower prices. Prices for the best cattle reached almost $10 per hundredweight in 1932 and ranged down to just $1.50 for poorer varieties.[27]

The quality of the meat only began with the animal that left the farm. Realizing its potential depended on proper execution of the slaughtering process and preparation of the meat for shipment to company branch houses. "The buyer's opinion of [meat's] value depends largely on its appearance," admonished a handbook on proper slaughtering techniques. The importance of the "dressed" carcass's quality in an industry that relied on hand labor was a constant drag on packers' efforts to speed production. Imperatives to move beef through the plants as fast as possible to maximize income were at odds with the

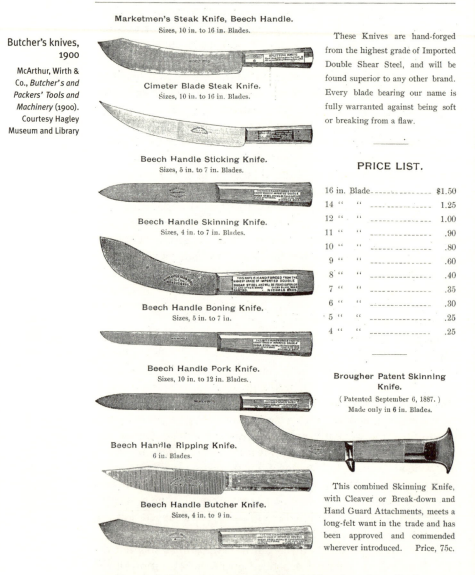

Marketmen's Steak Knife, Beech Handle.
Sizes, 10 in. to 16 in. Blades.

Cimeter Blade Steak Knife.
Sizes, 10 in. to 16 in. Blades.

These Knives are hand-forged from the highest grade of Imported Double Shear Steel, and will be found superior to any other brand. Every blade bearing our name is fully warranted against being soft or breaking from a flaw.

Beech Handle Sticking Knife.
Sizes, 5 in. to 7 in. Blades.

PRICE LIST.

16 in. Blade		$1.50
14 " "		1.25
12 " "		1.00
11 " "		.90
10 " "		.80
9 " "		.60
8 " "		.40
7 " "		.35
6 " "		.30
5 " "		.25
4 " "		.25

Beech Handle Skinning Knife.
Sizes, 4 in. to 7 in. Blades.

Beech Handle Boning Knife.
Sizes, 5 in. to 7 in.

Beech Handle Pork Knife.
Sizes, 10 in. to 12 in. Blades.

Brougher Patent Skinning Knife.
(Patented September 6, 1887.)
Made only in 6 in. Blades.

Beech Handle Ripping Knife.
6 in. Blades.

Beech Handle Butcher Knife.
Sizes, 4 in. to 9 in.

This combined Skinning Knife, with Cleaver or Break-down and Hand Guard Attachments, meets a long-felt want in the trade and has been approved and commended wherever introduced.   Price, 75c.

need to manufacture meat presentable for butcher shop display cases. Hence skilled butchers who could work with both speed and precision remained essential to providing beef for Americans' tables.[28]

Refrigeration freed beef production from its ties to local markets and allowed packers to fragment the work process, deskill labor, and introduce machinery in the more simple operations. Opportunities for larger volume led packing firms to transform the modest slaughtering sheds of 1870s Chicago into huge operations employing hundreds—sometimes thousands—of workers. To increase speed, firms experimented with methods designed to take "the work to the men instead of the men to the work," as Henry Ford observed upon seeing one Chicago operation. (Ford credited this visit with giving him the idea for the automobile assembly line.) Reorganized production systems and greater application of labor permitted vast increases in beef production. Yet production speed remained limited by the careful hand labor needed to prepare beef fit for the consumer marketplace.[29]

Beef sufficiently presentable for the butchers' display case began with the initial killing blow. The seemingly straightforward slitting of the animal's throat was not so simple as it might have appeared to visitors horrified at the blood-splattered man who did the sticking. Packinghouse supervisors were admonished to make sure that the sticker "severs both arteries and veins otherwise the beef will purge on the back and look discolored when finished." While cutting deeply enough to sever both blood vessels, the sticker also had to not cut so deeply as "to stick cattle through, for if this is done, when the bullock is thrown on its back, the blood flows onto the chime bones, causing a bad discoloration." Speed of production was balanced by precise hand labor performed under onerous conditions.[30]

Proper bleeding was part of the initial preparation of the beef for market. In the 1880s firms adopted the important innovation of using a hand windlass to hoist stunned cattle into the air before killing and "bleeding" them, a change from 1870s methods of "pumping" the animal's legs while it lay on the ground. Pumping the cattle to remove its blood produced beef "always of a dirty red color and it never looked right in the markets," recalled Neil Carbrey, a butcher from that era. Bleeding the cattle while elevated slowed the slaughtering process but improved the meat's color. The new sticking techniques emulated, to a certain extent, traditional Jewish killing methods whereby a *schoctim* cut the animal's throat then hung it up to bleed. "You can always distinguish their beef from the gentile beef," explained Carbrey, "as it is white and clean, with a beautiful live color to it, and it sells readily."[31]

A similar tension of speed and care marked the next production stage, removing the animal's hide. A mechanical hoist could be used to haul the shackled animal up to a chain that carried it to the sticker, but the modicum of continuous process came to a halt as the chain dropped the animal back down to the floor for the hide removal process. Turned into leather for industrial uses and personal consumption, the hide required careful attention in the removal process. The carcass had to rest on the ground while successive groups of butchers carefully cut into the "fell," the membrane separating the hide from the meat, so as to

avoid any nicks in the hide that could reduce its value or cuts into the flesh that would mar the meat cuts. Poor skinning would leave "a big black eye" on the meat, recalled Iowa kill floor worker Frank Hlavacek: "Boy you'd get hell for that." The hide removal process could neither be mechanized nor hurried, and it precluded a true continuous operation in the cattle "disassembly" line. Most plants tried to ease around this bottleneck by having multiple skinning beds, sometimes as many as eight or ten gangs, to increase production.[32]

Yet speed could not be ignored. A carcass remaining at the kill floor's 100° F temperature would spoil within thirty minutes. To do the job well and quickly, workers "used a knife in either hand." The floorsmen, unquestionably the most highly skilled workers in a slaughterhouse, skinned the cattle's belly and sides with a long-bladed knife. They had to leave the fell on the front shoulders, as otherwise the meat "shows black when coming from the cooler." Next butchers sawed open the breast, a task that required holding the saw "at the same angle at which the animal is laying" to avoid creating "a very bad looking brisket on one side of the beef." A hoist then lifted the rear legs off the ground so that the "rumper" could cut the hide away from the animal's rear end, taking "great care . . . not to get into the lean meat." A "backer" cut the hide away further, maneuvering his knife so that "the fat is not removed from the loin." Once the hide was fully removed and carcass gutted, splitters had to sever the backbone precisely down the middle so that the exposed bone exhibited the best color and look. Firms were admonished to "have choppers ground [to] different thicknesses for different boned cattle" so that the splitters, "who look out for themselves, generally speaking" would have the right tools to choose from.[33]

Once split the halves chilled for at least 24 hours so that natural enzyme processes would reverse the rigor mortis that set in immediately after killing and soften the meat. Before shipment the ribber who separated the halves in quarters made the final critical cuts. As the cut needed to be made in an area 1/2 inch wide, "it requires skilled work to do it rapidly and do it properly." Even the beef luggers who hefted the 250-pound quarters into railroad cars had to be careful. If not wedged tightly the "the swing and motion of the cars" would knock pieces together "and a bone from one quarter of meat will mangle and tear the meat on the quarter hanging next to it, often very seriously injuring its appearance."[34]

To kill and dress sixty cattle per hour in 1905 required a gang of seventy men, including four floorsmen, two splitters, and a half dozen rumpers, backers, and hide droppers. A small army of unskilled workers assisted these skilled men (and aspired to their positions), moving and positioning the meat, removing intestines, cleaning up the blood, washing the carcass, and so forth. Increasing production to eighty-four per hour took eighty-eight men, with most of the increase in labor coming at the skilled end while the unskilled labor needs remained about the same. Technological change over the next half century was minimal. "Backbreaking is what it was," recalled Louis Tickal about his work on the beef killing floors in the 1930s and 1940s. "This is all done by hand." Company equipment catalogs printed in the 1940s simply reproduced descriptive pictures of the beef slaughtering process from the 1920s because operations were largely the same. The principal innovation of a Cincinnati Butcher's Supply Company catalog that promised "revolutionized beef killing" was an electric-powered hoist that replaced human or animal powered windlasses previously used to raise and lower the cattle carcasses during the killing floor operation.[35]

Retail butchers operated in an entirely different environment than twentieth-century "slaughterhouse butchers." Their labor of creating consumer cuts out of the beef quarters took place in small shops isolated from their colleagues; packinghouse workers rubbed shoulders in

Retail beef cuts,
1920s

Evelyn G. Halliday
and Isabel T. Noble,
*Hows and Whys of
Cooking* (1928)

Arrangement of cuts similar to that in a chart by Armour & Co.

FIG. 45.—Wholesale cuts of beef, Chicago style

factories with hundreds, and often thousands, of fellow employees. While making beef quarters was standardized as much as possible, retail cuts were particularized by region and often fabricated on demand for consumer preference. In short, retail butchers took national beef and refined it for the local market place, much as Thomas De Voe had done in the nineteenth century.

Retail butchers had to know the different cuts characteristic of their region. "Nearly every city has its special ways and peculiarities of cutting meats," advised a packers' handbook. The New York "Kosher chuck" comprised one-third of the usable carcass weight while Chicago's version was but 28 percent of the animal. New York retail butchers also cut the most valuable pieces—the loin, ribs, and chuck—closer to the backbone than in Chicago, creating a larger flank, rump, and brisket plus an odd cut known as the "short hip . . . an abbreviated Chicago loin end plus a bit of the rump." Philadelphia and Boston cuts also varied significantly from the Chicago and New York styles.[36]

Retail butchers were refiners of consumer cuts and, at the same time, salesmen to women shoppers. These functions, though so different than the floorsmen's, conveyed a similar power in the procurement of beef. Given the "ease with which the meat cutter may move from one city to the other" noted a national study in 1928, they "are not particularly noted for constancy." All too often a meat shop owner would find his store "without a cutter, without previous notice." Like the packinghouses that tried to retain their skilled beef butchers, retail meat markets were advised to provide "constant employment, and a liberal wage" rather than follow a policy of "grinding down of their employees under low wages and poor working conditions."[37]

Within white ethnic and African American communities, slaughterhouse butchers and retail meat cutters were highly desirable occupations. According middle-class incomes despite its rough, blue-color character, butchering provided upward mobility for several generations of workers at the lower end of the American labor market. Internal promotion ladders, especially in the slaughterhouses, allowed workers without a high school degree to develop skill with a knife and rise to progressively better jobs as they showed their abilities to make meat in a manner that suited consumer expectations and company requirements.

Both groups of workers used their power over beef production to consolidate strong unions by the end of World War II. Beef kill workers led organizing efforts that created the United Packinghouse Workers of America, an organization affiliated with the Congress of Industrial Organizations. Befitting their economic location in a national industrial process, these workers focused on, and secured, national contracts with the major meatpacking firms by the end of World War II, standardizing wages and working conditions nationwide. Retail butchers enrolled in the Amalgamated Meat Cutters and Butchers Workmen, an organization affiliated with the American Federation of Labor. As their primary association and interests were in particular localities, the Amalgamated concentrated on regulating wages and hours within particular cities while accepting the variations from place to place. The capacity of butchers in slaughterhouses and retail stores to secure recognition, and contracts, came from their essential role: putting meat on the American table.

Beef did not become the dominant meat in America until the twentieth century. Centralized slaughtering operations and refrigeration in distribution and retailing were the twin portals to beef's entry into so many consumers' homes. By 1948, 88 percent of American families had at least some fresh beef in their diets, a larger proportion than any other meat variety. Perhaps even more dramatically, two-thirds of American families in 1965 were able to consume steaks. Jack London's parable had come to define the postwar success of the American beef industry.

This modernist success story carried with it a powerful cultural legacy from consumption practices in the pre–Civil War era of local beef. The hierarchy of beef cuts remained intact, with the porterhouse steak and rib roast at the top and tougher, leaner cuts at the bottom, albeit with chopped beef displacing stewing meat as the usable form of cheaper parts. Moreover, beef remained an anonymous product, with cuts distinguished by their origins in the animal carcass rather than the purveyor's name, despite the aspirations of beef producers to brand their products similar to pork.

Overcoming the problem of marketing beef in its fresh forms might have been attenuated, but the dilemma of meeting consumer expectations continued to hobble beef producers. As their nineteenth-century ancestors had, twentieth-century consumers expected their meat to be red and soft, and they time and again rejected efforts to retail frozen beef, either in packages or under clear packaging materials. Beef could not be transformed to the extent that other meat products could, mainly because consumer expectations were so stable and stubborn. It would be in the area of cured products, principally pork, where contemporary Americans would learn to eat meats that were fundamentally different from those served in their grandparents' homes.

Chapter Three

Pork

Laura Ingalls Wilder opens *Little House in the Big Woods,* her fictionalized recollections of childhood frontier experiences in the 1870s, with her family's food preparations for winter. "Butchering Time" filled a very special place in these preparations. In a cycle practiced by millions of rural American families from settlement well into the twentieth century, Wilder's homesteaders turned their pig into winter stores. The family had set the pig loose in the woods to forage on "nuts and acorns and roots" during the summer, penning it a few weeks before butchering time to fatten on corn and table scraps. When it came time, "Pa" and another (male) family member killed the hog, cleaned off the bristles in a scalding tub, hung it from a tree, removed the intestines, and left it to cool.

When the body heat had dissipated, the men divided the pig into its basic cuts—hams and front shoulders, sides, ribs, and belly—and salted them in preparation for curing in a brine solution and smoking. The choice spare ribs were saved to be eaten fresh at a special harvest dinner that evening. The smaller pieces then went to the women, who boiled the fat down into lard, turned the scraps into sausage meat, and stripped the head for the ingredients that made head cheese. When all was done, "The little house was fairly bursting with good food stored away for the long winter."[1]

As Wilder's story indicates, pork slaughtering and curing was widespread in rural America. A meat eaten cured rather than fresh, pork was America's preeminent meat before urban growth and home refrigeration made beef more accessible. In 1880 almost 50 percent of all pork was consumed in the form of ham or bacon, an amount larger than the entire national consumption of fresh and cured beef. Given the extensive (and often unreported) practice of home slaughtering, such as by Wilder's frontier family, these official census figures almost certainly understate the true level of Americans' pork consumption.

The close relationship between farm and kitchen influenced the categories of pork known to Americans before the Civil War, and formed the basis for the products of the commercial meat business. While farmers varied widely in their curing and cutting techniques, the emergence of a national—indeed international—market for meat encouraged standardization of processed pork varieties. In this way, vernacular preferences, reflecting the influences of national origin, geography, and climate, established the major categories of meat products.

The commercialization of pork transformed its treatment and preparation. National distribution networks of refrigerated cars expanded fresh pork's potential market at the same time as the new large meatpacking companies improved consumer-oriented forms of cured pork. Seeking to reduce the length of the curing process, firms introduced new chemicals and technologies into the production process, inciting public concern over the wholesomeness of products that once had been manufactured close to home. Popular revulsion against chemical additives triggered in large part the Meat Inspection Act and regulations of 1906, but this legislation was only a slight setback to pork's transformation in the twentieth century. New curing techniques revolutionized pork products at mid-century, as ham and bacon displaced barrel pork as the preeminent pork products in America.

The pork Americans ate in 1960 was a far different kind of meat than their grandparents had eaten. New technologies transformed popular conceptions of most pork products. Cured barrel pork disappeared, to be replaced by pork taken from the same areas of the pig, yet cut differently for contemporary dishes and eaten fresh, rather than cured. Bacon and ham did not look or taste as they had in Wilder's day; and while ham remained the elite pork product, bacon's place in the symbolic universe of American food was entirely new. Although pork remained a popular item for breakfast, lunch, and dinner, the meatpacking firms had utterly changed its form.

Country Pork

Wilder's story captures why pork was America's most widely consumed meat well into the twentieth century. The animal was well suited to farming families living near woodlands and growing grain. Pigs required little management and fattened easily on the leftovers of human consumption. They were easier than cattle for farmers to handle because their smaller size (120–250 pounds versus 800–1,000) made killing and curing far simpler. Butchering time naturally followed harvesting of field crops, as the weather turned cold and facilitated meat preservation. The flesh could be cured through materials easy to obtain and use—salt, sugar, and smoke—and thus stored for later consumption in an era without home refrigeration. Processing the pig fit into the family division of labor, with men performing the killing and cutting operations outside and women and children handling the meat that needed more detailed care to be turned into a usable product.

Pork was cured using traditional preservation methods handed down from well-established European techniques. Men first separated the "joints"—the ham (rear leg) and shoulder—from the body, and then cut the trunk into pieces small enough to place in a barrel filled with a brine solution, usually composed of salt, saltpeter, and sugar or molasses. Salt carried most of the curing burden, while sugar and molasses helped with taste and saltpeter improved the meat's color. Cured in this manner the meat could stay edible for six months to a year, as the extremely salty brine solution prevented bacteria from growing in the meat. James Fenimore Cooper acknowledged salt pork's place in American popular cuisine in *The Chainbearer* (1845), whose housewife considered "a family to be in a desperate way when the mother can see the bottom of the pork barrel." (Our contemporary aphorism, "scraping the bottom of the barrel" is a relic of this nineteenth-century notion.)

Complementing barrel pork were cuts cured with a combination of salt and smoke. Hams and shoulders often were dry cured—sprinkled with salt, saltpeter, and sugar rather than immersed in a liquid solution—then smoked to complete the curing process. Smoke added flavor, and the dry-cured meat could last well over a year without spoiling. Sometimes farmers separated the ribs from the pig's belly and cured the latter as bacon. Hams also were valuable commodities that could be exchanged for salt, sugar, or rum at local stores.

While families from all classes and regions offered pork for dinner, particular products were markers of America's social hierarchy. Barrel salt pork was a poor family's meat—shared by slaves, farmers, or wage earners—along with bacon, which was especially popular among rural southern whites. Elites favored hams and other choice dry-cured products in warm weather and fresh roasts during the colder months. They especially disdained barrel pork as "sea-junk," cured by "sopping in brine" that imparted a "villainous" and "nauseous" taste entirely different from the "savoury" dry salted pork varieties.[2]

While pork was America's leading meat in the nineteenth century, its cured products (aside from ham) occupied an inferior place in the meat hierarchy. Nineteenth-century cooking books treated salted barrel pork as an ingredient rather than an item deserving its own entry. "A little salted pork or bacon should always be kept in the house," advised Mary F. Henderson in *Practical Cooking and Dinner Giving*. "Pork makes a delicious flavoring for cooking other meats, and thin, small slices of breakfast bacon are a relishing garnish for beefsteak, veal cutlets, liver, etc." Many dishes whose titles featured beans or fish actually began with large quantities of bulk salt pork in the cooking pot. Boston baked beans or navy bean soup demanded a healthy chunk of salt pork, usually one-half pound of meat to every pint of beans. Slices of salt pork fried in a pan or kettle began recipes for chops or fish stews; breaded and fried salt pork served as a popular garnish for dishes in need of more taste. Beef roasts—especially from the leaner parts of the animal—were stitched through with "lardoons," rectangular chunks of salt pork (a half inch square and two to three inches long), to impart taste to the meat. Nineteenth-century cookbooks generally did not list salt pork among their recipes or in the index (even though many of their dishes included this meat), reflecting its invisibility despite high consumption levels, and the low status it occupied. Although she recommended many uses of salt pork, Henderson confessed "to having a decided prejudice against this meat."[3]

Elite prejudices aside, eighteenth- and nineteenth-century Americans highly valued and widely consumed wet-cured pork. Sarah F. McMahon's careful studies of probate inventories indicate the popularity of this form of meat. By the time of the American Revolution, 69 percent of her sample had barrel pork on hand in the summer, six to nine months after it had been prepared, and that proportion remained virtually constant through 1835. While poorer estates were less likely to have salt pork stores than more prosperous ones, nonetheless 50 percent of the estates valuing between $100 and $200 in 1835 contained salt pork, compared to 75 percent of the estates between $400 and $800. And it is noteworthy that this barreled meat was a significant enough asset to appear in probates at all.[4]

The pork dishes that commanded attention in cookbooks were for more exceptional parts of the pig, principally hams and fresh cuts, even though they were eaten far less frequently than barrel pork. Due to their size, hams were for special occasions, and the cookbooks indicated the respect that such dishes deserved. *The Century Cook Book* advised dotting a ham with pepper, then inserting a clove in the center of each spot of pepper and serving it with a vegetable cut into the shape of a rose to conceal the bone. Fresh pork remained a highly seasonal dish, unlike these cured products. Roasted or stewed pork generally was a fall or winter treat that accompanied curing the rest of the animal for later use. This cyclical consumption pattern, originating in climate's natural influences, assumed a cultural life of its own and induced suspicion among nineteenth-century Americans toward fresh pork eaten out of season. "By many, fresh pork is considered to be exceedingly unwholesome during the months of high temperature," Thomas De Voe wrote in 1866. "The instincts of experience no doubt lessen the demand for fresh pork during the heat of the summer trade." Wealthy Americans, whether living in cities with meat markets or on southern plantations, might be able to enjoy fresh pork year round. But for most nineteenth-century Americans, eating pork meant eating cured pork.[5]

Industrial Pork

Beginning with the rise of Cincinnati pork processing in the 1820s, entrepreneurs discovered that, whenever possible, it was cheaper to move the slaughterhouses and meat-processing facilities to the animals than to ship live animals to major population centers. So long as the meat could be kept from spoiling and transported economically, large-scale production facilities near livestock sources permitted economies of scale in meat production. Fragmenting the labor process to increase productivity and centralizing distribution reduced costs significantly—and eventually changed the types of pork consumed in America.

Climate and geography were the principal determinants of meat-packing's initial concentrations. Growth of internal transportation after 1815, principally roads, canals, and steamboats on existing water-ways, allowed commercial nodal points to emerge for packing cured meat, preeminently pork. Seizure of the Ohio valley from its native inhabitants in the early 1800s opened up large areas for corn cultivation, and as farming advanced so too did pig raising, with farmers essentially converting part of their corn crop to pork. They either slaughtered the pigs on the farm or drove herds to convenient river-front market centers for slaughter and processing. Initially general merchants along waterways entered pork packing as a minor adjunct of their business, but soon the steadily growing flow of pigs off local farms induced the first true meatpacking companies to emerge.

Its advantageous geographic location helped Cincinnati become America's leading antebellum pork-processing center. Perched on the Ohio River's banks in rich farming country, Cincinnati was a favorite destination for farmers eager to take advantage of its superior outlets to southern and eastern markets. The network of rivers, canals, and lakes available to Cincinnati shippers led an observer to comment that the city was located in "the centre of a circle which bears on the Atlantic in the east, the vast prairies on the west, the lake-counties on the north, and the Gulf of Mexico on the south." Annual production levels exceeded 100,000 hogs in the 1830s, then doubled to 250,000 in the 1840s and reached 400,000 on the eve of the Civil War. In the 1860–61 packing season, Cincinnati exported 70 million pounds of cured pork and bacon. By mid-century, Cincinnati's pork could be found on the tables of planters and slaves in the South and artisans and the upper class in East Coast cities. The town that grandly proclaimed itself the "Queen City" of the West achieved true fame with its more quotidian persona as the nation's "porkopolis."[6]

In the absence of reliable refrigeration, climate governed the cycle of meat operations in Cincinnati. Beginning in November, when the weather turned cold enough for animals to be slaughtered and chilled for curing, pigs clogged roads leading into town. When Frederick Law Olmsted departed from Cincinnati in November 1853, droves of hogs "filling the road from side to side for a long distance" impeded his progress. Roadside killing sheds located alongside or over streams on the city's outer fringes furiously slaughtered these animals during the cold months. These crude wood structures used the simple expedient of "movable lattice-work at the sides" to admit sufficient cold air to chill the carcasses.[7] Some pork quickly left the city as "green" or lightly cured meat, sent south on flatboats before the river froze.

Once cold weather paralyzed shipping, attention turned to curing pork in the brick buildings clustered along the Miami canal in Cincinnati's central industrial district. While hogs made their own way to the slaughterhouses on foot, their carcasses were trucked by the tens of thousands on wagons "piled up in rows as high as possible" through city streets to the packing plants. Pork packing was so ubiquitous during the winter that a visitor complained, "we could not look into a warehouse in the street without being agonized by the sight of thousands of dead corpses, heaped and piled upon one another." Once spring thawed the Ohio River, killing operations halted and shipments reached their peak as the meat companies desperately sought to sell their stock, acquired from farmers on bills of credit. Cured meat awaiting loading onto riverboats would "spread over the public landing, and block up every vacant space on the sidewalks, the public streets, and even adjacent lots otherwise vacant." By May the industry was in its slow season, only to surge once again in the fall with the harvest and return of cold weather.[8]

Cincinnati's pork packers were businessmen who rarely soiled their hands by actually cutting meat, unlike their entrepreneurial contemporaries in antebellum Manhattan such as Thomas De Voe. Rather than functioning in a daily market setting gauging sales through personal interactions with customers, Cincinnati's meat men gambled on long-term demand for pork products in distant ports and cities, anticipating that pigs purchased in November would be sold as bacon, ham, and lard six months later. They were more merchant than industrialist, better attuned to the vagaries of credit and demand for commodities than the mechanics of turning live animals into meat. Indeed, in the mid-1830s ten of Cincinnati's thirteen slaughterhouses employed the same superintendent, John W. Coleman, attesting to the attenuated participation of Cincinnati's pork packers in their own establishments' operations.[9]

The mercantile origin of Cincinnati's pork packers suggests why one visitor to Cincinnati in the 1850s observed, "every third person you meet is unquestionably a pig merchant." The meat business in this antebellum center was highly decentralized, as "in most cases the business of curing pork is separate from the slaughtering," and it was relatively easy to establish small curing and packing operations. In 1864 over seventy firms slaughtered or cured meat in Cincinnati, ranging from individual proprietors like James Magill, who advertised as a general purpose "PORK AND BEEF PACKER AND A COMMISSION MERCHANT," to established pork packing firm S. Davis Jr. and Co., which boasted that its "Diamond Brand" family ham had won a silver medal at the 1850 Ohio State Fair. Davis, one of Cincinnati's largest firms, handled 18,000 hogs in the late 1840s, less than 5 percent of Cincinnati's output, leaving plenty of room for small merchants to enter the business so long as they had the credit to turn hogs into pork and the connections to sell the meat in southern or eastern markets.[10]

By the late 1850s Chicago was challenging Cincinnati as the nation's leading pork-packing center. The expansion of the nation's rail network explains much of this change, along with the continued westward movement of agriculture. As railroad track mileage grew to 9,000 in 1850 and 31,000 by 1860, canals and rivers became less desirable means for transporting meat. Railroads had two principal virtues in comparison to water transport: trunk routes could convey food to eastern markets on a year-round basis, and feeder lines could enter the countryside and bring livestock from landlocked farms directly to central markets. Located astride this rail network, Chicago took full advantage of its transportation advantage and passed Cincinnati as the nation's leading meatpacking center during the Civil War. By 1870 Chicago produced cured pork products valued at $19 million, twice as much as Cincinnati. Chicago held its leadership position as "hog butcher for the world" for the next hundred years.[11]

A principal architect of Chicago's rise, Philip Danforth Armour, was a packer in the tradition of Cincinnati's "pig merchants." An entrepreneur whose initial capital came from selling supplies to California gold prospectors, Armour entered the meat business during the Civil War by establishing a partnership with an experienced butcher and packer, John Plankinton, in Milwaukee. While Armour knew little about killing and processing meat, he was an experienced commodities trader. In 1864 he made a huge profit on cured pork by anticipating a Northern victory and signing commitments to sell pork to the Union army at $40 a barrel rather than hold out for higher prices. When the market collapsed and pork fell to $18 a barrel, Armour and Plankinton made approximately $1 million simply by fulfilling their contracts with pork purchased at lower prices on the open market. These profits financed Armour's expansion into Chicago, as well as other promising new packing centers such as Kansas City.

Armour was the biggest player in Chicago's burgeoning post–Civil War meatpacking industry, centered at the Union Stockyards in the city's southwest side. Census data conveys the qualitatively larger scale of Chicago's meatpacking operations. In 1880 Chicago's slaughtering and meatpacking plants averaged slightly over a hundred employees and produced meat worth $85 million, while Cincinnati's operations employed but twenty-three workers apiece and generated an aggregate value of $11 million. By 1890 Armour alone killed more pigs than the combined total of all Cincinnati packers.[12]

Perceptually, Chicago's stockyards district also occupied a very different space in the city's economy and the imagination of observers. Unlike antebellum New York and Cincinnati, Chicago's meat factories were located six miles southwest of the downtown and isolated from adjacent neighborhoods by polluted streams and acres of railroad tracks. Rather than being an integral part of the city's life, the stockyard district was an otherworldly spectacle, with the sounds and sight of animals awaiting their death, the mysterious hustle and bustle of thousands of men engaged in the meat business, huge ominous slaughterhouses towering over the yards, and their ever-present odor. Visitors to Cincinnati and New York were dismayed at the omnipresent meat business; in Chicago the stockyards became a popular destination for tourists seeking to marvel (and cringe) at the modernism of the industry's production methods. Rudyard Kipling granted that "every Englishman" who came to Chicago would visit the yards to marvel at the sight of "a township of cattle pens" that stretched "as far as the eye can reach." Entering a "death factory" to see the killing process left an indelible impression on Kipling, as it must have for other visitors. "There was no place for hand or foot that was not coated with the thickness of dried blood," he recalled, "and the stench of it in the nostrils bred fear."[13]

Meat's perishable nature still restricted the industry in the decade after the Civil War; it was Chicago that turned meatpacking into a year-round business. Until freshly slaughtered meat could be kept from spoiling during warm weather, the enterprise remained imprisoned by the seasons. In 1870 over 90 percent of Chicago's pork products were processed during the late fall and winter months. With the expansion of production and improved rail lines, however, both the interest and feasibility of using bulk ice refrigeration increased. By the mid-1870s Armour and other large Chicago packers had invested in cold storage facilities cooled during the summer with ice from the Great Lakes. Now able to use artificial means to chill the animal after slaughter, summer hog processing increased steadily throughout the 1870s from a half million animals in 1873 to over four million in 1880. With its productivity augmented by year-round operations, Chicago accounted for over one-third of all meat produced in the United States in 1890.[14]

The concentration of hog processing in Cincinnati after 1820 and then Chicago following the Civil War opened up opportunities for innovations in production methods. A greater division of labor, rather than new technology, allowed for increased productivity. In a typical 1830s slaughtering shed on Cincinnati's outskirts, a gang of twenty men killed 620 hogs in eight hours. One felled the animals with a two-pointed hammer while standing on their backs in a tightly packed pen. An observer was reminded "of a man on a raft of logs, the logs rather loosely secured, and no one sufficient to support him," as the striker was forced "to save himself" by stepping from one pig to another.[15] The animals were then dragged to butchers who cut their throats, and rolled them into a vat of hot water to loosen their hair. The carcasses were pushed through the water by long sticks and then pulled out by a pair of brawny men who passed them along to a team of butchers using sharp knives to remove hair and bristles. In one plant the scraping table consisted of sixteen or more men each with narrowly defined tasks for shaving the animal. Two removed bristles "worth saving for the brush makers," then passed it on to men who shaved hair off one side, and in turn passed the carcass along to men who attacked the other side. Workers then hung the hairless carcass from its rear legs so that a final team of butchers could remove the hogs' entrails. It then cooled for twenty-four hours before transfer to the separate curing and packing operations.

When Frederick Law Olmsted visited Cincinnati in the 1850s, he avoided the slaughterhouses, "satisfied at seeing the rivers of blood that flowed from them." But he did enter a cutting and curing operation and there witnessed a similar application of division of labor, rather than machinery, to speed production. He observed a "human chopping-machine" consisting of no more than a "plank table, two men to lift and turn, two to wield the cleavers." The efficiency of these men was such that "no iron cog-wheels could work with more regular motion." As butchers separated the pig carcass into parts, "attendants, aided by trucks and dumbwaiters, dispatch each to its separate destiny," the curing cellars below where the pork was preserved before shipment. Olmsted timed these men dismembering a hog every 35 seconds; with several parallel stations in operation, a cutting and processing plant such as this could pack fifteen to twenty thousand hogs during the winter season.[16]

By the time Philip Armour entered the pork business and helped turn Chicago into the "hog butcher for the world," the basic structure of hog killing operations was well established. The higher productivity of the Chicago hog plants rested in year-round production (made possible by refrigeration), incremental technological innovation, and the consolidation of continuous process production. Engineers continued to improve the "disassembly" line by using machines to eliminate bottlenecks. Turn-of-the-century plants sent hog shacklers directly into pens where they fastened chains to the live animal's leg. Once attached the chain pulled the pig into the air and conveyed it to the "killers" who severed the animal's jugular vein. After the animal's blood drained, the

Cutting up hogs
by hand,
mid-nineteenth
century

*Seven Days in
Chicago* (1876).
Courtesy of Smith-
sonian Institution
Libraries,
Washington, DC

chain dragged the carcass through the scalding tub. A large metal grate replaced the men who had once removed the hog from the water. Technology was also used to perform the tedious (and labor-intensive) hair-removal process. A machine with a cone-shaped aperture surrounded by knives mounted on adjustable springs shaved the carcass while it remained on the chain. As in so many American industries, simplification of tasks preceded, and made possible, mechanization of production stages.

In cutting operations, slaughterhouses increased productivity 25 percent by fragmenting butcher's tasks and installing a rotating wheel or conveyer track to bring the animal to workers performing specific cuts. When the hog reached the end of the de-hairing table, a stick called a "gambril" inserted between the rear legs suspended the animal from a wheel ten feet in diameter that rotated parallel to the floor. The hogs were hung on hooks about four feet apart on the outer edge, and then the wheel turned, bringing the work to the butchers automatically. Such ease of movement permitted a greater division of labor, such as having one man remove the large intestines and another handle the rest of the innards, each having but twelve seconds for his task. In a few years the revolving wheel was replaced by an overhead chain system that moved the hog carcass. It was simpler and accomplished the same end: allowing a greater division of labor by moving the animal from one man to the next.[17]

In the early twentieth century, power-driven devices such as bandsaws and conveyer belts sped operations, but slaughtering and butchering remained overwhelmingly a hand operation. Closely observing work in a 1904 plant, labor economist John R. Commons astutely observed, "skill has become specialized to fit the anatomy." The irregular size of the animals and variations in the most desirable angles for cutting meat rendered human skill with the knife indispensable and baffled inventors seeking to automate cutting operations. The killing stroke had to be delivered properly so that the animal would finish bleeding before entering the scalding tub. Splitters split the carcass in two with a seven-pound cleaver that had to cut precisely through the middle of the backbone. "Headers" needed expert eye-hand coordination and considerable physical strength to slice the animal's backbone at the neck and yet leave some skin attached between the head and body—and to do so for one animal every ten seconds! While a conveyer might move the carcass from one butcher to another, it was still separated into pieces by men wielding knives who took years to become fast and proficient in their tasks.

One early-twentieth-century plant employed 156 men to slaughter and "dress" 400–450 hogs an hour. Job titles such as "ham facer," "lard puller," and "pluck and paunch trimmer" convey the work's manual character. In 1905 the best-paid jobs were the stickers, splitters, and ham trimmers at 35 cents per hour. Butchers such as cheekers and belly

Pig Scraping Machine—Capacity 8 Pigs per minute.

Automated hair-removal machine for pigs, late nineteenth century

Douglas's Encyclopaedia (1907). Courtesy of Smithsonian Institution Libraries, Washington, DC

trimmers held a series of positions paying 22 cents to 32½ cents. Ranging down to 15 cents an hour were various laboring positions entailing strength and use of hands but not a knife.[18] By the 1940s some butchers used cutting devices such as a front foot saw rather than a handheld knife, clearly a decline in skill requirements. Better organization of plants also reduced the actual number of laboring positions as conveyer belts took the place of men moving meat from one place to another. Overall, however, hog-cutting operations in the late 1940s were virtually as labor intensive as in 1905.

Industrialization of hog processing systematized the categories of pork consumed by Americans. Vernacular styles became regulated varieties, as boards of trade in meatpacking centers sought consistency in the terms applied to meat sold to the public. In the nineteenth century particular care was devoted to the categories of barreled pork (containing at least 196 pounds in each barrel), the chief product of these packing centers and the variety most vulnerable to bastardization and deception.

By the 1830s, regulations of Cincinnati's Board of Trade delineated among barrel pork products and in so doing facilitated selling these goods to distinct (and distant) markets. "Clear pork" comprised the best class among barreled meat as it included the sides of large hogs, with the ribs cut out. It was destined for New England residents who "in the line of pickled pork, buy nothing short of the best."[19] "Mess pork" came next in quality (and price) and included two rumps as well as the sides. The navy and commercial marine took a great deal of mess pork. A barrel of "prime pork," one step down, typically contained sides from lighter hogs along with two shoulders and two jowls. It was marketed for maritime "ship use" as well as "the southern market," where it generally ended up on plantations. The lowest grade was "bulk pork," which could contain any part of the hog (including the head and feet) and usually was "sent off in flat boats to the lower Mississippi" where it probably ended up distributed out of New Orleans to slaves in the Deep South.[20]

Distinction among barrel pork varieties grew along with the centralization of the industry in Chicago. By the 1880s the Chicago Board of Trade had doubled the number of barrel pork categories to eight. The elite clear pork, for example, now had a regular variety with the backbone and half of the ribs removed from the sides of "extra-heavy, well-fatted hogs," and an "extra clear pork" version that was completely boned and limited to fourteen pieces to the barrel. Prime pork metamorphosed into regular prime, extra prime, and prime mess. Only regular prime could contain heads, and prime mess meat had to be in four-pound square chunks and packed so that for every twenty shoulder pieces there were thirty from the side. Cures also were clearly specified, with mess pork defined as containing "not less than forty pounds of Turk's Island, St. Ustes, or Trepanne', or 45 pounds of foreign and domestic course salt" and cut into strips "not exceeding six and one half inches wide."[21]

Industrializing common country products made pork widely available throughout nineteenth-century America. Consumer habits bracketed the innovations of manufacturers so strongly that the new large meatpacking firms largely reproduced traditional products, albeit in far larger quantities. But industrialization also triggered an expanding spiral of change among pork products, as firms sought ways to improve cures and reduce costs at the same time as consumers developed different expectations for the foods they preferred to eat. In time industrialized pork would change the types of pork Americans favored.

Remaking Modern Pork

At the same time as they were improving production technologies and systematizing barrel pork varieties, the national packing companies were seeking more profitable venues for sales of pig meat. They devoted considerable energy to expanding consumption of branded pork meats, principally ham and bacon. While the barrel pork purchased from local merchants had no brand name, bacon and ham did. As the leading items in a packer's portfolio, "the quality and reputation" of these two varieties established the firm's "prestige" in the consumer marketplace and stimulated sales of its more prosaic products. As the receptacle of human efforts to refashion meat products, processed meats provided a means by which firms could distinguish their goods. Through their branding strategy meatpacking companies transformed the cured pork consumed by Americans.[22]

Increasing demand for bacon and ham relative to other cured meats was apparent by the end of the nineteenth century. Total pork consumption grew from 2.84 billion pounds in 1880 to 4.3 billion in 1900, while cured products remained stable at 70 percent of total pork consumption. Within the cured pork category, however, bacon and ham were growing far more rapidly. Between 1890 and 1900, production of these items increased 48 percent while barrel pork grew just 9 percent. Astute packing executives could see that anonymous barrel pork was "a part of the packing house business which is becoming of less importance year by year" and that the future, for pork packers, was in branded bacon and ham.[23]

Pursuing a branding strategy, however, was hindered by turn-of-the-century production and curing methods. The packers could take the pig from the sticker to the coolers in fifteen minutes, but trimming the meat remained hand labor, and, worse yet, curing times were unchanged from what they had been a hundred years before. A 1905 manual for prospective meatpacking entrepreneurs warned that going into commercial porkpacking was far more expensive than entering the beef trade, as "in slaughtering hogs fully 70 percent of the carcass goes into curing departments, to stay there anywhere from thirty to ninety days."[24] Hence, beginning in the 1880s and continuing almost without rest subsequently, packing firms tinkered with the physical chemistry of curing materials in order to speed production of branded pork products.

While salt, saltpeter, spices, and smoke were used universally to cure ham and bacon, altering the proportion of ingredients could produce significant variations in taste. Firms capitalized on the specificity of curing techniques by branding their hams and bacon with standardized recipes. As late as the 1890s cures varied among different Swift plants. "The cures for pork were all secret," recalled Louis Swift, son of the firm's founder. "The head man at each plant had his secret formula. By paying him a large salary, we obtained his services." Spoilage of meat at the Kansas City plant prompted the company's founder to terminate this practice. In a story that may be apocryphal, Gustavus Swift called together his plant managers and ordered them to turn over their recipes to him. He chose the best and decreed that all plants would use that combination of ingredients.[25]

The focal points for innovation, bacon and ham, occupied different places in the pork hierarchy. Hams were unquestionably "the finest parts of the animal" because of the meat's quality and the careful curing process. Fine differences among hams could be established by the foods fed to the pigs, the cut and trim of the raw meat, and the ingredients in the curing solution. Smithfield ham, for example, came from hogs that pastured in the peanut fields of Virginia and the Carolinas after the harvest. To earn the sobriquet "Smithfield ham" they had to be dry cured, smoked over a hickory fire, then rubbed with black molasses and black peppers and hung to dry for at least one year. A guidebook for hotels advised, "they improve with age up to three years, which is considered a prime age." To legally "brand" this ham against imitation, Virginia producers secured legislative passage of a statute specifying that Smithfield hams came from "the carcasses of peanut-fed hogs . . . and which are cured, treated, smoked and processed in the town of Smithfield, in the state of Virginia."[26]

Most other varieties of nineteenth-century ham were fully cured in a liquid brine solution after an initial dry-curing period. Consistently, though, hams were subsequently dried and smoked similar to bacon. Such heavy curing led to claims that hams "may be kept any length of time." Stories of hams upward of fifty years old remaining wholesome and tasty abound in nineteenth-century sources.[27]

Bacon, by contrast, was a common meat, perhaps slightly better than bulk pork but still a rough, nonelite provision. "You'r greasy and salty and smokey as sin/But of all grub we love you the best," a poem simply titled "Bacon" opened. The tension between bacon's nutritional features and status as a food item persisted throughout the poem. "You'r as good in December as May/You have always come in when the fresh meat has ceased" the poem gushed; as a result, "the rough 'course of empire' was greased/By the bacon we fried on the way." But while rhapsodizing bacon's role as provision, the poem also expressed a troubled undercurrent about its symbolic associations. "We've sworn you were not fit for white men to eat," the poem admitted, and "called you by names that I dare not repeat."[28]

Consumed throughout the country, bacon was especially popular in the South where local production was widespread. One British traveler to Kentucky was disgusted at the penchant for southerners to "eat salt meat three times a day" and disparagingly termed bacon "the favourite diet of all the inhabitants of the State." Clearly a low-status meat, bacon remained widely popular because, preserved through dry-cure methods, it lasted longer than salt pork soaked in brine and could be purchased in smaller quantities. And while travelers might have tired of its "greasy and salty and smokey" taste, it certainly was better than low-grade barrel pork.[29]

Contradictory definitions of what constituted bacon reflected the incomplete confluence of vernacular tastes with the commercializing meat economy. Although Cincinnati and Chicago packers defined bacon as cured pigs' bellies, in traditional country parlance bacon referred to the dry-curing method of impregnating the meat with salt followed by smoking and drying. Virtually any part of the pig could become bacon if cured in this manner. "The bacon cured here is not to be equaled in any part of the world, their hams in particular," expounded a proud Virginian on the state's curing methods around the time of the American Revolution. After first applying sugar to the meat, the bacon/ham would "lie in salt for 10 days or a fortnight." Rubbed by saltpeter or hickory ashes to impart a red color, the meat dried for three to four weeks over a slow smoky fire using "nothing but hickory wood." Preparing the meat in this manner preserved it "for several years." Whether this process referred to the joints, belly, or sides was unimportant—what mattered was use of the dry-cure method.[30]

This prosaic country dry-cure method, developed from rural experience and not scientific knowledge, nonetheless was highly effective and subsequently endorsed by twentieth-century physical chemistry. Imparting sugar to the meat drew out the water, magnifying salt's dehydrating effects. This initial treatment ensured that harmful bacteria did not gain a foothold in the critical period immediately after slaughtering, as bacteria need water and oxygen within which to grow. Such success carried a price, unfortunately. Deprival of oxygen turned meat an unappetizing grayish color, as it prevented myoglobin proteins from "blooming" and turning red. Saltpeter counteracted this effect by giving meat (albeit artificially) "the ruddy appearance always desired."[31] Smoking not only dried the meat but also (and especially in the case of hickory wood) produced acids with their own preservative effects.

Industrial pork refined, but did not transform, dry-curing methods. Large-scale production for national commerce encouraged defining bacon more precisely as cured pig's bellies, but the only initial change was the scale of curing operations. Bacon went into huge wooden (later metal) containers (holding 800–900 pounds) to cure, with layers of salt, sugar, and saltpeter sprinkled between them. Smokehouses grew larger and relied on steam heat and hickory sawdust rather than a slowly burning fire to impart flavor and preservations, but the long curing times did not change appreciably. A 1905 meatpacking handbook recommended holding bacon in dry salt for thirty-five days and then smoking it for an additional twenty, a week longer than the traditional methods of colonial Virginia.[32]

Dry-curing techniques were not convenient for high-volume bacon production. Adding too much sugar to the cure negatively affected the meat's color, as it darkened during cooking. Yet failing to add enough sugar left the bacon tasting extremely salty, and lacking "from an American standpoint, at least, the flavor which is obtained in sweet pickle bacon." A liquid "sweet pickle" solution, in addition to more reliably imparting the right flavor, also reduced handling and material costs. Thousands of pounds could be wet cured in the 1,500-gallon vats that were standard packinghouse equipment in the early twentieth century. Once the meat went to the smokehouse the remaining liquid could be reused for the next batch.[33]

With all these advantages little wonder that packing firms began to cure bacon with wet methods similar to hams. By the early twentieth century only "fancy" bacon was made through dry-cure methods. In so doing the firms turned the definition of this form of pork upside down; rather than referring to meat cured in a particular way, bacon now referred to a part of the pig, regardless of how it was cured.

Firms also tried to make the curing process more reliable by adding new ingredients, principally boracic acid and borax. Both were antiseptic compounds that inhibited bacterial growth but did not actually cure meat. Employed to address a range of ailments in the nineteenth century, they were especially useful in wartime to block infections in soldiers' wounds. Their attraction to meatpacking firms is obvious; they promised to prevent food decay and thereby facilitate conveying "meat, hams, bacon, etc., from very distant places to market in a perfectly sweet and fresh state."[34] A curing agent marketed by B. Heller that doubtless contained these materials (but did not indicate so) promised that chopped meat treated with the compound would "retain its fresh, appetizing appearance from one to three weeks if necessary, without the use of ice or cold storage." The Preservaline Company sold a similar compound that they promised "WILL KEEP FRESH pork and liver sausage when exposed on your counter, and in the hottest weather, for at least one week." Neither of these manufacturers admitted that boracic acid and borax allowed them to make these disturbing claims, although Preservaline had by that time acquired "extensive borax mines" in California, whose product doubtless went into its curing products.[35]

Although the first use of these chemicals in curing solutions is murky, turn-of-the-century industrial cures routinely contained substantial quantities. A 1905 sweet pickle solution for ham called for 3,750 pounds of salt, 700 pounds of sugar, 160 pounds of saltpeter, and 50 pounds of borax to be dissolved in a 1,500-gallon vat. In dry cures, borax was routinely part of the curing agent and also sprinkled on prior to shipping, "to prevent [the meat] turning slippery or moldy." As there were neither product labeling requirements nor federal regulation of domestic meat products, inclusion of these ingredients was not evident to consumers.[36]

These two ingredients were at the center of the public health controversy that finally resulted, after publication of Upton Sinclair's exposé *The Jungle,* in federal regulation of the meat industry. Defenders of boracic acid and borax attributed criticism of these ingredients to the way "every innovation and improvement of this kind has always first to contend with a considerable amount of unreasoning prejudice." Rather than being "injurious" to consumers, these materials were an improvement over older curing methods and were of "inestimable value" in safeguarding the public's health. Supporters were able to marshal considerable medical opinion in support of these claims, including tests showing that the human body passed boracic acid and borax quite rapidly.[37]

Critics, led by Harvey Wiley, a chemist and leading exponent of government regulation of the meat industry, countered by stressing the incremental impact of these products. Employing human subjects, Wiley conducted careful chemical tests for the Department of Agriculture demonstrating, he alleged, that over time these agents interfered with "the processes of digestion and absorption" and could damage the kidneys. Wiley's tests would not have convinced medical experts today; he only used twelve volunteers, all white men in their twenties, and his results were not consistent. He won, however, the political point that the danger posed by cumulative ingestion of these ingredients, especially among children and the elderly, was a sufficient basis to prohibit them. Tapping into the outrage inspired by Sinclair's novel, he argued that borax and boracic acid were dangerous both in their own right and because they permitted preserving meat "with very much less care, in a very much shorter time, and at a very greatly reduced expense." In 1906, as part of the regulations governing the meatpacking industry, the Department of Agriculture banned all meat additives other than those specifically authorized, which were limited to "common salt, sugar, wood smoke, vinegar, pure spices, and pending further inquiry, saltpeter."[38]

Banning unproven additives such as boracic acid did not spread to saltpeter, an ingredient used for so long without harm that even Wiley's ministrations failed to stir government action. But meat industry firms continued to search for ways to shorten the two- to three-month curing period for ham and bacon that saddled them with large meat inventories. In the 1910s Preservaline and a new firm, Griffith Laboratories, began investigating the use of sodium nitrite and nitrate to speed curing times. Griffith began importing nitrite under the brand name "Prague Salt" in 1925 and secured USDA permission to include it in curing solutions the same year.

To apply the cure more efficiently, Griffith adapted ham "pumping" methods that had originated in the nineteenth century. "Stitch" pumping (as it was known) entailed using a needle inserted by hand into the meat in several places to introduce the brine solution. Stitch pumping had limited effectiveness, primarily because the hand methods made it hard to control delivery of the brine. Packers were warned not to pump so hard as "to burst the tissue," but at the same time to make sure enough curing agent seeped around the bone, "as this is where decomposition usually sets in." For these reasons firms employed stitch pumping only for their cheaper hams and bacon, while "the special, or leader brands are cured without pumping." Packers especially worried that creating holes in the ham increased chances of bacterial contamination and spoilage.[39]

Griffith's innovation was to move from "stitch" to "artery" or vein pumping which introduced the cure through the circulatory system rather than directly into the flesh. Vein pumping greatly magnified nitrite's impact on curing times; an industry survey of this innovation later noted that the advantages of the new curing agents, "and the ability to obtain a mildly-flavored ham with the greatest speed, were not fully realized until the advent of artery pumping." Many years later Griffith could gloat that its mid-1920s advertisements that made "astounding claims" of curing hams in just five to ten days had proven correct.[40]

Jesse Vaughn, a packinghouse worker who started at a small Chicago plant in 1924, lived through these dramatic changes in curing tech-

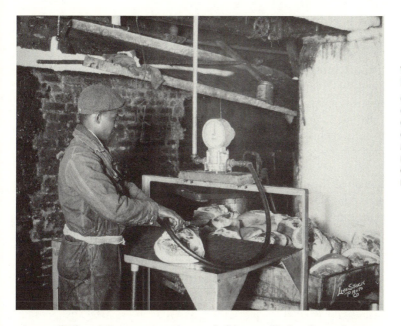

niques. "When I first went to the yards," he recalled to "get a ham on the street, it took 90 days." More than that, the curing process was labor intensive. The ham started in the vat "face up," but after thirty days the pickle had to be changed and the ham moved "over on its face." One month later the pickle was changed again, and "the ham [put] back up on its back. And reason for that, that pickle took that long to go through the ham to cure it." But Vaughn explained that once "they learned how to vein pump those hams," curing times dropped dramatically to just five days.[41]

Firms like Armour trumpeted faster cures as the advance of modernism. "The difference between the hams of a generation ago and the Star Hams turned out today by Armour and Company" was, the company thundered, "almost as vast as the difference between kerosene light and electric light." These advertisements ignored, not surprisingly, the negative consequences on the quality of nitrite-cured, "vein-pumped" hams. "It required so much pickle in that ham, the packer would over pump that 10 percent," recalled Vaughn. Hams cured this way could turn out to be "nothing but water." Endorsing Vaughn's shop floor knowledge, chemical supply firm B. Heller and Company admitted that adding so much liquid could have an "adverse effect upon flavor and appearance."[42]

Fast-cure methods also changed hams' taste as they did not permit sufficient time for the meat to age and develop a rich flavor. Griffith Laboratories president Carroll Griffith admitted that the fast curing process did not produce the same "rich ripe flavor that came with the long-time cure." With fruit, "the optimum point is somewhere between the green fruit and the over-ripe fruit," he explained by analogy. "A similar breakdown occurs in the aging of meat products." Curing manufacturers claimed that improved mixtures in the 1950s (though identical in composition to their 1930s predecessor) would "recapture the distinctive, fully cured flavor associated with old-time cures." The continued appeal of "old-time" hams suggests they were never able to fully replicate the flavor of hams cured with nineteenth-century methods.[43]

Nitrite and vein pumping improved ham-curing times, but packers remained frustrated at the production end of the process. No matter how hard they tried, ham boning and skinning resisted mechanization. Knives and cutting braces may have improved slightly, but for over a century following the Civil War, preparing the ham for curing still entailed a group of men standing around a table doing the work by hand. "I think we had about 10 guys in the gang," recalled Vaughn. "Number 1, we had what we'd call skinners that would take the ham off, and put the ham back on the table, and it goes to the next guy, he would take what they call the 'aitch' bone out . . . After the guy take the aitch bone out of the ham, he puts it back on the table and the next guy take the ham off the table and he bone it, and when he puts it back on the table it's completely boned." Great care had to be taken with the knife to cut off enough fat to give the ham a "clean" look but not so much to interfere with the cure or reduce the weight. To become a true "first class ham boner," Vaughn recalled, would "take quite some time." Befitting their necessary and highly skilled labor, ham boners were the best-paid workers in the pork side of modern meatpacking operations. A gang such as Vaughn's could process 200 to 300 hams an hour.[44]

Skilled ham boners enjoyed job security and opportunities similar to their associates in beef butchering operations. As the value of the hams owed enormously to their appearance, men skilled in removing bones and trimming fat held a strong position in the labor market. Men like Jesse Vaughn commanded top wages in the packinghouse and never feared unemployment, even in the depths of the Great Depression.

While cutting methods changed only slightly over a century, the hams were not at all the same. Wet cured and less inoculated with salt than their predecessors, hams became softer and sweeter—or blander, depending on taste preferences. Easier and quicker to refine, they were a highly profitable item. But their elite place in the pork hierarchy did not change. Hams remained a special occasion meat, advertised generally in association with the Thanksgiving-Christmas-Easter holiday cycle.

Modernizing bacon took a different path. Nitrites slightly improved the curing process's reliability, but there was no dramatic advance in

THE HAM PART of ham and eggs is carefully prepared for curing on another Rex equipped conveyor line.

ON THE WAY TO PORK

curing times similar to hams until the 1950s. Instead, bacon went through a second transformation principally through improved manufacturing technology. In so doing bacon broke with its past associations as a poor person's meat to become one consumed widely at all socioeconomic levels.

Until the 1910s bacon universally left the packinghouse as cured slabs of four to ten pounds. "Packaging" was limited to wrapping in waxed paper embossed with the firm's name. Consumers either could pay extra for bacon sliced for them by the butchers, or, as in most cases, slice the slab as needed at home.

Beginning in 1915, packing firms began to incorporate the slicing and wrapping operations into bacon production. In so doing they were accommodating themselves to consumer preferences for a more manageable form of this meat rather than addressing a problem posed from a production standpoint, such as long curing times, as they had done with ham. Creating a bacon slicing and packaging operation entailed substantial capital investment and employment of more workers, and longer retention of the product in the packinghouse. (By contrast, improvements in ham curing had dramatically reduced the time between purchase and sale of the meat.) The return, packers hoped, was higher value added with the sliced bacon and, above all, upgrading the reputation of this meat so that it would no longer carry its association as a lower-class meat. Since they could not speed it out of the packing houses any more quickly, packers tried to expand demand for the meat and to "upscale" its clientele.

Branding cured pork in Swift & Company advertisement, c. 1900

Warshaw Collection of Business Americana—Meat. Courtesy of Archives Center, National Museum of American History, Behring Center, Smithsonian Institution

Late 1920s Armour advertisements expressed these messages. One set of advertisements promoted "The NEW and MODERN Armour and Company," which had solved the problem of wet-cured bacon having a poor texture through its new "double-f" curing process. Doubtless referring to use of nitrite in the curing solution, Armour claimed to have made its Star Bacon a product suitable not only for breakfasts but also in "fancy luncheon dishes, smart supper menus." Another series of advertisements placed in magazines like *McCall's* and *Women's Home Companion* featured fashionably dressed upscale men and women consuming Star Bacon purchased "in the new pound or half-pound cartons." Modern curing technology and convenience would, Armour hoped, expand the demand for bacon among the urban, white middle class.[45]

The nitrite curing techniques for bacon only entailed converting dry-curing areas to wet cures, but getting bacon into convenient containers was far more complicated. The first bacon-slicing machinery went into operation around 1915, and could make about a hundred slices per minute that were "absolutely uniform in thickness" to facilitate cooking. Once sliced, the rest of the operation was by hand. Women stacked, weighed, and wrapped the bacon for shipment to retail meat shops. By the time of Armour's late 1920s advertising campaign, speeds had improved to 300 slices per minute for chilled bacon. The cured belly had to be inserted into the cutter by hand, manually pushed into the cutting knife, then pulled out from the hopper by hand and given to the bacon line, which continued to rely on hand labor to sort, weigh, and wrap the meat. By the late 1940s technology had advanced to the point of creating a more continuous process. Bacon slicers that automatically fed the cured belly into the cutting blade could generate 800 slices per minute. A conveyer belt pulled the slices to shingling and wrapping machines, though operators were still necessary to weigh and position the strips.[46]

As bacon slicing and wrapping equipment became more elaborate, however, proper preparation of the meat became even more important—and capital and labor intensive. Hog bellies that naturally came in different sizes had to be standardized for the new technology to operate properly. Pieces that were too thick or too wide jammed the machines; if they were too warm (above 38° F) the cut pieces would be mashed rather than neatly sliced. Along the hog cutting line, where previously bellies simply had to be separated from the back and ribs before going into the curing tanks, firms now had to add bacon trimmers to properly size the meat, a belly roller to press it to consistent thickness, and a press that "perfectly squared and universally sized" bellies so they could be handled by the slicing machines. A cutting room handling 550 hogs per hour in the 1940s required six men to trim the belly and one to operate the belly press, so that the meat would be acceptable to the bacon slicing equipment. In the specialized buildings constructed for the new bacon operations, several lines of around ten women each prepared the consumer-ready bacon. The convenient sliced bacon package snapped up by food shoppers required a great deal of investment, machinery, and labor to prepare.[47]

Bacon finally began to move more quickly through packinghouses in the 1950s as firms developed equipment injecting curing solution through dozens of small needles—updated "stitch pumping" methods in essence. The "PerMEATor" built by the Cincinnati Boss Company allowed a continuous flow of graded bellies to automatically slide into the machine. "In one rhythmic, elliptical motion" the needle-laden header completed its task "of penetrating, injecting the cure and moving the belly along." Twenty-four hours after passing through the machine, the curing solution had permeated the bellies and they were ready for the smoke house. A 1952 version could handle 240 bellies an hour; speeds doubled within a few years.[48]

The quick cures, however, were ineffective against a new problem, the deterioration that set in when the slices were placed under a clear wrapper that admitted light and oxygen. The unsliced bacon slab had naturally precluded damage to its interior; exposing the bacon by slicing it created a new set of problems. "All bacon can and does begin to lose flavor and change color from the moment it comes off the slicer," admitted the head of Armour's Research Division in 1955. It might take a "fine palate" to notice any deterioration one or two days after slicing and packaging, "but after 7 to 10 days, most of us could readily detect a difference." Better packaging was the answer to this new problem.[49]

Firms began using cellophane to package sliced bacon in the late 1920s, almost simultaneous with the material's introduction to the United States by DuPont. Cellophane became a far better covering after 1945 when DuPont developed a version that permitted enough evaporation to keep the meat "not too dry, with consequent excessive shrinkage and discoloration, and not too wet, with consequent mold and bacterial deterioration." Cellophane of this type may have filled the "requirements of consumer appeal, convenience and economy," but it was ineffective in a key production requirement: keeping oxygen away from the meat and thereby lengthening "the period of full-flavor and color retention."[50]

Vacuum packaging of sliced bacon (and other meats) emerged in the late 1950s to address this problem, drawing on cellophane varieties and new films, principally polyvinyl chloride compounds, to accomplish these ends and to permit bacon to be both convenient and durable. Featuring a new package designed by Raymond Loewy, Armour established more elaborate vacuum packaging operations that entailed additional

"BOSS" Bacon Squaring Press—No. 220

Bacon squaring press standardizing dimensions "to permit ready slicing"

Cincinnati Butchers Supply Co., *Boss Abattoirs and Packing Plant Equipment*, Catalog 54. Cincinnati Boss Company Records. Courtesy of Archives Center, National Museum of American History, Behring Center, Smithsonian Institution

This is a powerful machine that will form bacons to a determined length, breadth and thickness to permit ready slicing and avoid all waste of end or uneven pieces.

The press is fitted with centrifugal pump which is positive in action and works quickly and with precision.

The open bacon mold is 11 in. wide, 26½ in. long, and the squared bacons are pressed to 7, 8, 9 or 10 in. widths by 16 to 26 in. lengths.

The press will handle 160 to 180 bacons per hour when operated by one man, while two men can handle 260 to 280 bacons. This is based on a capacity of 80 to 90 bacons per hour for one slicing unit.

One man operating the press only, is able to serve two slicing units. If three slicing units are to be supplied by the bacon press, two men are necessary for operating the press.

Bacons that are properly chilled, are shaped without any danger of separating the lean from the fat and retain the squared form for an indefinite time.

Floor space, 7'0" x 2'8".
Height over all, 7'2".
Domestic weight, 4700 lbs.
Export weight, 5500 lbs.
Cubic feet, 260.
Code: ZIJEG.

labor and capital investment needs, and which extended the shelf life of refrigerated bacon to six weeks.

In stark contrast to firms' efforts to reduce costs associated with producing ham, companies invested fortunes in the facilities, technology, and workers needed to produce appealing sliced bacon. In doing so in the 1920s and 1930s, however, they assigned workers to jobs based on assumptions about the appropriate occupations for men and women, as well as the permissible roles for African Americans in the labor process. A strict racial and sexual division of labor characterized sliced bacon operations. Men dominated the belly cutting and trimming positions that took place in the hog cut room prior to curing. These new tasks went to men as they were an extension of what had been traditionally male butchering jobs. Men also performed the key step in the production of bacon, slicing the cured pig's belly by a machine, placing them in control of the production speeds and also earning the highest rate of pay in the sliced bacon department. White women, on the other hand, dominated the sliced bacon packaging operation, arranging the slices in rows, weighing, and wrapping.[51]

Sliced bacon departments performed an unusual dual function in packinghouses, as places to advertise the company's products and to prepare them for distribution. Visitors to packinghouses routinely went through the sliced bacon room, and some departments even had one glass wall to facilitate observation by tours. Firms consciously constructed what visitors saw: clean, well-lit rooms in which neatly dressed white women performed their tasks while seated comfortably at long tables.

Limiting employment in sliced bacon to white women reflected the department's combined role as producer and promoter of this product. In the late 1920s three-fifths of sliced bacon workers were under the age of twenty-five, and 84 percent were native-born. The presence of young, white "American" women was well suited to a positive public representation of meat production. Employers excluded black women from sliced bacon for similar reasons. One management representative explained to a curious observer that the "picture" presented in the department would "seem prettier to visitors if it were all white." Another said bluntly that "only white hands are fit to touch the meat" and that the public would not buy the firms' products "if they know colored women were employed." While there was no evidence that the millions of Americans who purchased bacon paid much attention to the production line, these management attitudes indicate a larger point. Placing black women in sliced bacon work might have interfered with the "upscaling" strategy with which they hoped to expand consumption of sliced bacon.[52]

As sliced bacon production expanded in the 1950s, these departments became centers of a campaign by the principal meatpacking union, the United Packinghouse Workers of America (UPWA), to extend the employment options of black women. Under the seniority and nondiscrimination clauses of union contracts, the UPWA succeeded in opening employment in this area to black women by the late 1950s. This major advance for black women's employment opportunities had none of the consequences projected by the companies when they had excluded black women in the 1920s. Consumers were far more interested in the ease of using sliced bacon and did not exhibit any worries over who placed it in the packages.[53]

Mid-1960s consumption data provides a measure of pork producers' accomplishments. Although bacon was more expensive than round steak and pork chops in 1920, at 65 cents per pound in 1960 it was 20 cents cheaper than the former and 40 cents less than the latter. Bacon had been remade both as a form of meat and in its relation to other meats. Bacon consumption remained higher in the South than the Northeast, but it was a true national meat whose appeal crossed income levels. Slightly more than 60 percent of all U.S. families purchased bacon, with higher income groups slightly more likely to do so than lower. The "upscaling" strategy had worked, even as the retail price drop had made bacon cheaper to purchase.[54]

Producers were far less successful with ham. Its price relative to beef cuts did not change significantly between 1900 and 1965, remaining slightly cheaper than round steak and more expensive than chuck roasts and hamburger meat. Correspondingly, ham retained its place as a relatively exclusive pork product. Slightly over 30 percent of all households purchased hams, and upper-income groups were more likely to obtain it than were families in lower brackets. The principal change was greater consumption at the lowest income level, reflecting increased postwar purchasing power and the popularity of the ham sandwich! While ham's form and processing methods had changed, the meat remained at the apex of the cured pork hierarchy.[55]

Conclusion: Pork Remade

By the 1960s, pork was an entirely different meat than it had been a century before. Barrel salt pork, America's preeminent meat in the eighteenth and nineteenth centuries, was utterly absent from the 1963 Census of Manufactures and virtually disappeared from home consumption even at the lowest income levels. Sliced bacon, which did not appear as a census category at all before World War II, was produced in equal volume to cured hams in 1963, for a total cured pork production of 3 billion pounds.[56] The decline of salt pork probably was a good thing for the American diet. Whatever the problems of curing agents such as nitrites, the fatty, salty pork (cured at a ratio of one pound salt to four pounds meat) that typified eighteenth- and nineteenth-century meat consumption doubtless was not good for the heart or for digestion.

Changes among cured products accompanied a sharp rise in fresh pork consumption. Cuts that had previously been cured, or available fresh only at certain times of the year, were widely obtainable by the 1960s. Consumption surveys in 1965 documented that fresh cuts had risen to close to 50 percent of pork consumption across regions and income levels, a major change in the form that pork was reaching the dining table.

Fresh pork rose from the ashes of barrel pork's demise. With the packing firms' refrigerated distribution network providing alternatives to preserving pork in salty brine solutions, pork processors began to think of pig meat in different forms and shapes. Technology, in this

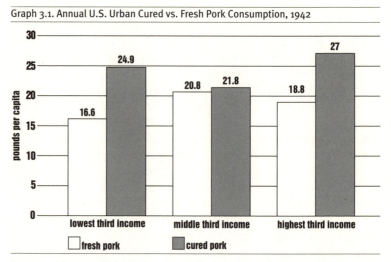

Graph 3.1. Annual U.S. Urban Cured vs. Fresh Pork Consumption, 1942

Sources: U.S. Department of Agriculture, "Family Food Consumption in the United States, Spring 1942," Miscellaneous Publication no. 550 (1942), esp. 82, 88, 98; U.S. Department of Agriculture, Agricultural Research Service, "Food Consumption of Households in the United States" (spring 1965).
Note: See Graphs, chap. 1.

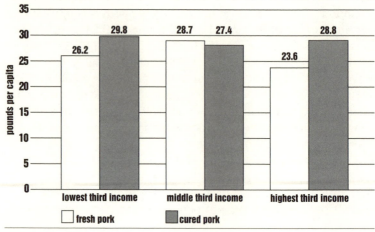

Graph 3.2. Annual U.S. Urban Cured vs. Fresh Pork Consumption, 1965

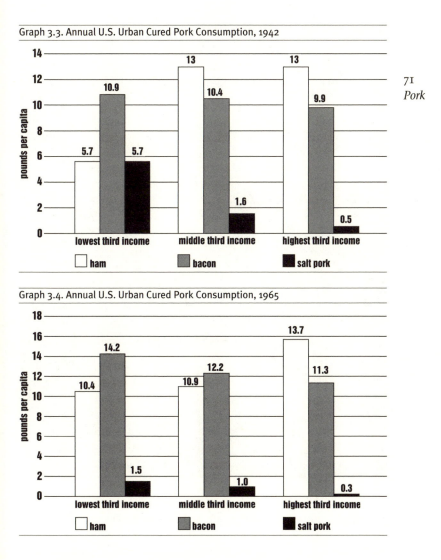

Graph 3.3. Annual U.S. Urban Cured Pork Consumption, 1942

Graph 3.4. Annual U.S. Urban Cured Pork Consumption, 1965

case, expanded the conceptual menu of pork manufacturers. In the 1880s one packing company "discovered" (the term used by contemporary Neal Carbrey) "the now well-known primal parts of pork cuts." With different notions of how to subdivide the animal came development of new cutting devices. Invention of the long, two-handled loin-pulling knife took place around 1900 as the market for fresh pork loin was taking off. The knife's shape suggests its shop floor origins as butchers sought easier ways to fabricate the new, fresh products. It "was modeled after the cooper's drawing knife and bent to the shape of the loin," recalled Carbrey, and reduced the loin removal time by a factor of five. Not surprisingly, pork chops became widely available year

Pork cuts, 1930.
Note racial and
regional associa-
tions with certain
pork cuts

Fortune, February
1930. Courtesy
Hagley Museum and
Library

This ear may someday be a sausage. But it is possible to make a silk purse out of a sow's ear

From this, cutlets

Pickled snouts are good

Boston butts are much esteemed in Philadelphia

The British like this part very much

Darkies are very fond of the tail. A curling tail means a happy, healthy pig.

It takes two pig-skins to make a suitcase

Southerners like salt pork

Chop Chop Chop Chop Chop Chop Chop Chop Chop

Spare ribs here

This is called the "picnic." No-one knows why

About one-eighth of a hog is bacon

This is the ham

Best of all, darkies like chitterlings

Southwestern pigs, fed on nuts, have spindly shanks

Pigs knuckles must have sauerkraut. Most knuckles are from forelegs

round around this time, as the loin (from which they are cut) no longer ended up in barrels of salt pork and instead could be kept fresh in packing firm branch houses and butcher shop chill rooms.[57]

The rise of fresh pork, and the highly processed and prepared varieties like bacon and boned ham, augmented hand labor's importance in pork processing. Technological innovation largely entered where prior refinement of the division of labor permitted introduction of a machine to perform narrow, repetitive, and automatic operations. Automatic bacon-wrapping machinery replaced female operatives, for example. Yet the irregular size of the animals and cuts and the care necessary to handle the meat properly so that it remains wholesome remain impediments to a truly continuous production line. Men (and now women) wielding knives (sometimes power operated) still kill and cut up the pigs, which insist on entering the packinghouses at slightly different weights and sizes. The loin puller uses a long two-handled knife that has changed very little in a century. Ham boners remain essential to production and one of the most highly paid workers in a packinghouse. One hundred years after John R. Commons observed workers in Chicago's plants, tools and skill are still specialized to fit the pig's anatomy.

The transformation of pork comprised quite different trajectories of change. Consumer preferences were the primary inducement for product innovations leading to sliced bacon. Creating convenient, ready-to-cook bacon in the packinghouse was a source of a great deal of trouble for meatpacking companies. Expanding consumer demand for bacon, rather than reducing costs or streamlining production methods, largely accounts for firms' efforts to refine this product.

Ham processing innovations, however, cannot be traced to consumer preferences. Here, the incentive for change came from firms' efforts to reduce production costs, especially the expensive retention of this product during the three-month curing process. Artery pumping was all about speedier cures and concomitantly lower costs, not better taste. Introducing borax and boracic acid improved reliability of the cures for shipping purposes (for both hams and bacon) thereby reducing spoilage, and not to make cured pork taste any better.

In both cases, however, the origin of product changes was not a sufficient explanation for the actual process of innovation. Efforts to create a consumer-oriented sliced bacon were shaped and constrained by the meat's physical properties. New bacon production methods were dually constituted out of efforts to make the product attractive to middle-class consumers and to adapt extant technologies and preservation methods. In so doing, meatpacking firms created distinctively American bacon, which a delegation of British meatpackers observed in 1951 "bears little or no resemblance to bacon as it is known in Britain." American bacon came from the belly and comprised but 20 percent of the carcass weight, "in comparison with British method whereby in producing Wiltshire sides 79% of the dead-weight is converted into bacon." American consumers may have wanted convenient sliced bacon, but it was not their choice that bacon in the United States took the particular form that it did. Production imperatives, rather than consumer preferences, gave us our particular form of modern, convenient, bacon.[58]

Conversely, consumer preferences constrained firms' efforts to reduce ham production costs. Initially stitch pumping created an inferior product, forcing firms to accept longer curing time in order to retain positive brand recognition value for their hams. Then the fast cure methods of the 1930s, while saving firms a great deal of money, met with heavy criticism for creating bland water-logged hams. The soft-boned hams even had to be forced into otherwise useless "ham shaped" containers so that they looked like hams to consumers! Government regulations forced firms to limit the water added to ham, and curing agent suppliers tried to refine their products in order to restore traditional cured taste. Consumer preferences were such that the ham could not be remade, quite unlike the fate of sliced bacon.

Pork's natural features remained a defining barrier for companies' efforts to market their meat. Transformed pork remained a meat of solid pieces, albeit trimmed, squashed, and sometimes sliced. Standardizing pork was hard so long as whole pieces were the source of the cuts eaten in America. The step beyond curing meats was to refashion their contours into a malleable package whose size could be standardized—a process that has given us the hot dog.

Chapter Four

Hot Dogs

The 1939 New York City World's Fair, known especially for its claim to portray the "World of Tomorrow," featured daring architecture and pathbreaking consumer goods such as the first public displays of television and nylon stockings. To highlight as its principal example of the "food of tomorrow," the Swift meatpacking company selected the simple hot dog. Swift housed its exhibit in a building resembling a super-airliner, which presumably would whisk visitors away to see the foods that would be served in the future. Inside, large crowds peered into glass-paneled rooms to marvel at the latest methods in hot dog manufacturing techniques.

The casual visitors were not the only ones fascinated by Swift's display. The meatpacking industry's principal journal, the *National Provisioner,* carried a long illustrated article on the exhibition. It admonished progressive meatpackers to visit the company's pavilion so they could "study carefully" the "new equipment and methods" Swift employed, "which may bring about fundamental changes in some plant operating procedures." The most "novel and interesting" innovation, according to the journal, "is that meat is not handled manually in any of the machines."[1]

To the *National Provisioner,* Swift's breakthrough with its "food of tomorrow" was fully automating the labor-intensive job of making hot dogs and creating, in essence, a continuous flow process. This innovation, boasted the journal, allowed millions of world's fair visitors to consume the products Swift made so efficiently and wholesomely. Technological innovation and modernism thus went hand in hand with providing nutritious and easy-to-prepare food for all Americans.

Swift applied its innovative production methods to one of human beings' most enduring forms of meat. Hot dogs are a variety of sausage, and sausage represents a solution by many cultures to meat's troublesome properties—its perishability and in animals of irregular size. (Even the ancient Egyptians ate sausage!) The normal course of butchering produced a great deal of meat attached to parts of the animal not suited to cooking or curing as whole pieces. Sausages functioned as a package to reconstitute meat from different parts of the animal as products of similar sizes. The meat could be put into man-made containers or, more generally, packed in the tubular intestines of animals, more benignly known as casings. Since the meat was being, in essence, recreated from scraps, sausage became an area of wide creativity, combining different types of meats and spices and employing a range of preservation techniques.

Swift's bold display of hot dog manufacturing techniques trumpeted the arrival of a new American meat. Neither beef nor pork nor chicken, but a meat cocktail defined not by its contents but instead by its size, shape, and curative qualities, hot dogs were a ubiquitous food item by the 1960s. In the first half of the century hot dogs became a "fun" food associated with enjoying commercial recreation at amusement parks, the seashore, and sporting events. Charlie Brown's aphorism that "A hot dog tastes better with a baseball game in front of it" indicates the symbolic association of this food with fun and play.

In the 1950s and 1960s hot dogs entered the home and sales soared. Easy to keep in the refrigerators that had become prevalent by the 1950s and well suited for children, hot dogs fit well into the lives of the baby boom generation. The irritating yet impossible to ignore Oscar Mayer ditty expressed manufacturers' hopes that "everyone would like me" when it came to hot dogs. Indeed 1960s statistics indicated this was the case, as the "average" American ate seventy-five hot dogs each year.

The emergence of this new meat came from a complex interaction between the manufacturing needs of large meat companies and the preferences of Americans seeking a quick and easy meal. Meatpacking firms expanded sausage and meat byproduct output in the late nineteenth century as they sought to profit from animal parts previously given or thrown away. In so doing they learned from food purveyors that there were enormous sales opportunities for easy-to-eat sausage varieties. Dramatic production innovations ensued, as firms standardized and accelerated frankfurter manufacturing. Following World War II, food processors seized opportunities for increased home consumption of branded hot dogs by taking advantage of existing consumer attitudes toward this product, especially its aura as a fun meat. Building on the technological changes portended by Swift's World's Fair exhibit, firms completed the process of turning hot dog meat into a cheap meat that needed only a minimum of human labor to emerge as ready-to-eat wieners.

Country Sausage and Urban Red Hots

County sausage was widespread in preindustrial America, as it had been in rural societies for thousands of years. Farmwomen typically had as their special task during "butchering time" salvaging scraps from slaughtered animals for family consumption. It was a seasonal, cold weather activity that persisted wherever country butchering took place.

Martha Ballard noted that one cold December day in 1803 she had been "trying hog lard; cut sauches meat." Then one January day she had "cleaned the skins for sausage." Widespread sales of sausage choppers across the American countryside in the nineteenth century augmented country sausage production. *Hints to Young Housekeepers* published in 1878 advised preparing for sausage-making parties by having "plenty of pies and breads baked . . . so that every member of the family, that is able, may devote herself to the work." (It evidently was unthinkable for male household members to participate in this activity.) One hundred pounds of sausage could be culled from 1,200 pounds of pork "by trimming off every part that can be spared."[2]

Sausages prepared in this manner could be eaten relatively soon or cured to last until next fall's harvest. For short-term storage, sausage mixtures were packed into stone pans or glass jars, covered with lard, and kept in a cool place. Longer-term curing required mixing salt and saltpeter into the chopped meat, stuffing the combination into animal intestines, soaking in a brine solution for several weeks, then smoking alongside the hams and bacons. "Young Housekeepers" were advised that cured in this way "sausage will keep all summer."[3]

While reconstituting sausage in casings facilitated curing, it added a particularly onerous task to butchering time: cleaning of the intestines. "If skins are used, they cannot be prepared with too much care" advised *The Great Economical Tea Co. Cook Book*. Catherine Beecher conveys just how tricky this was. The casings had to be emptied of any partially digested food, washed, and cut into six-foot lengths. Then Beecher advised using a candle rod to turn the segment inside out, so the interior could be washed "very thoroughly" and scraped before soaking in salt water in preparation for stuffing. "It is a very difficult job to scrape them clean without tearing them," warned Beecher. Little wonder that farm women heeded the advice that sausage "are about as well made into cakes."[4]

Rural sausage recipes indicate a great deal of variation in their components. Local tastes and the vicissitudes of the harvest precluded consistency in composition. Often the recipes did not distinguish between types of meat, only specifying proportions such as 22 pounds of meat (two-thirds lean and one-third fat), a half-pound of salt, three tablespoons of sage and pepper, and two of thyme. Form and curing methods, however, were far more consistently noted and associated with particular varieties. Bologna sausage, the only sausage consistently identified by name in mid-nineteenth-century cookbooks, could have any mixture of beef, pork, or veal. It was distinguished by stuffing the meat "into large skins" (usually beef intestines) and cured with a combination of heat and smoke. Doubtless there were countless varieties of sausage in rural American society, reflecting local meat supplies and spice preferences, and recipes may have changed as availability of certain products waxed and waned.[5]

The commercial urban meat economy of the mid-nineteenth century encouraged systematization of these home-prepared varieties. In Thomas De Voe's antebellum New York City, sausages were prepared by men specializing in that trade, who collected the intestines and other offal from the killing sheds and prepared the meat in their own establishments. "Blood puddings" and "common puddings" were forms of sausage using scraps of head meat, internal organs, and sometimes blood, stuffed into pig or beef intestines, cooked, and then sold cheaply in urban markets to "the poorer classes, and especially the hungry laborer" for 3 or 4 cents a pound. Sausages containing these ingredients also could be dry cured like bacon for later use, as could intestines for sale as chitterlings. In post–Civil War cities, vendors peddled these products (along with bread to hold them) to urban working men on lunch breaks and to revelers in beer gardens and pubs.[6]

By the late nineteenth century, commercial markets could offer dozens of sausage varieties, their numbers augmented by types brought to America by immigrants. Italian varieties are prominent in the 1911 *Grocer's Encyclopedia,* such as Mortadelli and Salami, along with French versions, including Lyonnaise and Parisian. These elite sausages, with precisely detailed meat and spice compositions, competed on the market with common types like blood sausage for which a German-influenced recipe advised, "use all kinds of cheek meat, heart, lungs and pork rinds. It is hardly necessary to give any proportion for the mixing of these, as the quantity can be regulated entirely by the material at hand." Bologna had split into a classy imported variety (mostly from Germany, not Italy) and an American version that included "such small pieces of meat as cannot be used for any other purpose" such as beef hearts. Both types remained bologna as they were stuffed into beef casings and cured with smoke.[7]

With so many varieties on the market, and produced with ingredients that might vary each week, curing methods defined the three principal sausage categories. Fresh sausage, usually pork but often incorporating inferior beef trimmings, was not cured at all and intended for immediate consumption. While intended for sale soon after preparation, its contents could include trimmings purchased "when they are cheap," salted and held for a few months "until the prices are high" for sausage products. A second sausage category included the lightly cured and cooked varieties that could keep a few days or weeks depending on how they were stored. Fresh pork sausage treated in this way to prolong its shelf life and varieties such as bologna were typical examples. "Summer sausage" comprised the third category; it was distinguished from others by a drying and smoking process so that "this sausage will keep

Sheet music, c.1900. Note the typical water-filled, heated container (closed) that keeps the hot dogs warm and the box with bread rolls.

for months without being cooked if properly handled." While there is no clear evidence why "summer" became a term applied to dried sausage, it probably stems from the meat's rural origins. Prepared and cured during the late fall butchering cycle, "summer sausage" suggested how long it could safely be consumed, rather than any particular meat content.[8]

Hot dogs (then known as frankfurters or wieners) first appear in late-nineteenth-century recipe books. They were based on traditions imported from either Frankfurt or Vienna (Wien). Turn-of-the-century recipes defined frankfurters as lightly cured and smoked sausages containing finely chopped pork, beef, sometimes veal, and always plenty of fat. The contents could include, among other ingredients, beef cheek meat, cooked tripe (stomach), pork kidneys, and beef, pork, and veal trimmings. The meat would be cured with salt and chemicals before going into the casings, smoked for a few hours to add flavor, and then precooked so they were ready to eat when sold to consumers. Composition and flavoring varied widely in these recipe books; consistently, however, they directed that the meat should be stuffed into sheep or narrow hog casings less than three-quarters of an inch thick and "linked" or divided into segments five to ten inches long.

Its size and manufacturing method would give hot dogs their special destiny among American sausage varieties. Similar to bologna as a relatively soft, lightly cured and cooked sausage variety, frankfurters were smaller than their thicker cousin. They were far more tender than dried sausage and easier to prepare than raw sausage as they were cooked before leaving the factory. Intended for immediate consumption, "as they readily become dry and unpalatable," they did not have to be sliced like bologna or summer sausage, or cooked for as long as fresh sausage. From this inadvertent constellation of advantages would come hot dogs' great success.[9]

By the early twentieth century, frankfurters were beginning to stand out among other sausage varieties because of their popularity at sites of popular entertainment. The first Coney Island hot dog stands went into business in the 1870s, and frankfurters pleased crowds at the 1893 Columbian exposition in Chicago. They became a standard food item at late-nineteenth-century baseball parks, and an apocryphal story has the moniker "hot dog" becoming synonymous with frankfurters when stadium vendors began shouting "get your red hot dogs here." (There is, however, no clear documentation for this incident cited so often as the source of the term "hot dog." Like many food terms its true source remains a mystery.) By the 1910 and 1920s, hot dogs were a fixture at sites of popular culture.

These hot dog vendors functioned not just as food purveyors but also as mediators between meat producers and the consumer market. Out of their self interest in commercial advancement they promoted increased consumption of this product. One ambitious employee of Coney Island's famous Feltman's restaurant noticed its popularity among young revelers and used hot dogs to set up his own food business. Jewish immigrant Nathan Handwerker started his own stand in 1916 (calling it after his first name) and fought his way into the Coney Island market by cutting his hot dog prices to a nickel. He also branded Nathan's hot dogs with a proprietary recipe developed by his wife. When the New York subway established a Coney Island station in the early 1920s Nathan's business surged, as its huge signs promoting five-cent franks loomed at riders disembarking to head for beachside attractions. Little wonder that the 1931 song "It's Hot Dog Time at Coney"

declared that "With Lemonade refreshing and mustard for dressing, the Hot Dog rules the Ocean side."

Simple convenience and economy explain much of hot dogs' appeal. Cooked before leaving the packinghouse, they only needed to be heated in water or lightly seared on a grill to bring out their flavor. The thin, bread-wrapped frankfurter fit easily into the mouths of men and women eager to enjoy themselves and not waste time dining or worry about the food falling on their clothes. Frankfurters also were cheap, giving them a democratic aura fitted for plebeian popular culture. "It wasn't on the Island of Capri/It wasn't on the beach at Waikiki/It wasn't at a perfume counter in Paree where we met," declared a 1939 song performed by Sammy Kaye and his orchestra. But "a spark of love ignited" and "The bubbles in our soda pop/Just hit us like champagne" when the two met by chance "at a little Hot Dog stand."[10]

In making this transition to a ubiquitous item at amusement parks, sporting events, and beachside stands, the German frankfurter transmuted into the American hot dog. While Cincinnati's wandering "Vienna sausage man" of the 1870s seemed a peculiar representative of the town's German district, the vendors who made it "hot dog time at Coney" had no ethnic connotations. Hot dogs were so Americanized that at the same time as the 1939 World's Fair, President Franklin Roosevelt served them to King George VI of England as "America's most typical food." (Unnoticed was the irony of serving a sausage once known as the frankfurter to the leader of a nation threatened by Germany.) On the eve of World War II, hot dogs had crossed over from being an ethnic food to an American food.

Unseen behind the hot dog carts and boardwalk grills were dramatic innovations in production methods that facilitated and stimulated consumer desire for hot dogs. The vendors needed a consistent product to secure a regular clientele in the highly competitive sites of urban amusements. "Branding" their hot dogs was not possible unless they could be assured of receiving a product of predictable size, consistency, and taste from the food manufacturer. In one legendary story, Oscar Mayer (the firm's founder) blocked shipment of a frankfurter batch to a Chicago baseball stadium because it did not meet the team's usual standards. Because of their close relationship with consumers, vendors approved of meat processing firms' efforts to develop systematic production methods that would provide hot dogs with consistent flavor and texture.

Coincident with food retailers' search for reliable hot dogs was a dilemma meat firms faced on the production side. As firms expanded beef and pork output in the late nineteenth century, better utilization of meat trimmings, internal organs, and other byproducts became a pressing issue. Rather than dispose of these materials, the national meatpacking companies decided to turn them into profits by adding the necessary technology and personnel to their factories. Lard, glue, hides, grease, and fertilizer were among the new product lines that the large meatpacking firms introduced around 1900 as they widened the scope of their manufacturing operations. Thus was born the saying that meatpackers used everything but the oink and the moo from the animals they slaughtered.

Sausages constituted one of the most profitable byproduct branches. Plenty of meat was left over after cattle quarters went out the door and pork cuts went into curing containers. Internal organs, meat trimmings along bones and from the head were, as one former Swift executive claimed, "as wholesome as porterhouse steak, but not so palatable." The "art of sausage making," in his judgment, consisted of "taking these low priced products and making from them a palatable, whole-some and at the same time economical article."[11]

Firms certainly mastered this art after 1900. As meat production grew significantly in the early twentieth century, the rate of increase for processed products such as sausage outpaced traditional fresh cuts. Between 1904 and 1925, fresh meat production grew by 66 percent, to 9.5 billion pounds. In the same period, sausage production by packing-houses almost tripled, to 900 million pounds. Accomplishing such augmented production entailed revolutionizing nineteenth-century sausage manufacturing methods.[12]

When they established the web of operations necessary to make cheap hot dogs, meatpacking firms embedded a gendered division of labor in the new work process, much as they had in bacon manufactur-ing. Until the late nineteenth century, commercial meat processing was overwhelmingly men's work; the 1890 census records show barely a thousand women in this industry. By 1920 women comprised about 15 percent of the packinghouse workforce, overwhelmingly concen-trated in sausage and canning operations. Moreover, women's employ-ment was in turn distinguished by racial and ethnic characteristics. Young, native-born, white women worked in the showcase sliced ba-con department while ethnic and black women—often married and with children—predominated in sausage-preparation operations. Ap-pearance may have mattered in selling sliced bacon, but reliability was primary in departments generally unseen by the public. And these were women who could be counted on to show up for work because their wages contributed significantly to family income.[13]

Employers exclusively hired women into departments that had been either transformed or created in response to the growth of the mass consumer market for processed products. Even though manage-ment paid women less than men, female substitution only occurred in areas undergoing substantial changes in production methods, such as sausage. Management either resisted, or perhaps never considered, in-troducing women into what had been male departments devoted to the production of fresh meat.

To prepare sausage meat, firms created trimming departments around 1900 staffed entirely by female workers. After the men carved the livestock into basic cuts, these women separated fat or rind from the lean in small pieces of leftover meat. Once trimmed, the meat could be sent for curing and stored, or dispatched immediately toward sausage manufacturing or canning. The department's structure symbolically encoded women's subordinate relationship to male workers; men in the pork cut department upstairs pushed their scraps down chutes that dropped them on the tables of the female trimmers.

Skill and work intensification, rather than machinery, generated high productivity. Because there was no chain or sequential work process to stimulate speed, firms paid women for each piece they trimmed, with their total earnings based on the quantity of lean meat accumulated daily. Working on almost frozen meat in dingy rooms kept to near freezing temperatures, women used long-handled forks to pull meat from under the chutes toward them, which led to occasional scuffles over particularly desirable lean scraps. "One [would] reach over here for a big piece of meat that had a lot of lean in it," recalled former Chicago Swift worker Philip Weightman, "another one would walk over and start fighting with the hook to get it."[14]

Good pork trimmers took up to a year to learn their craft and could earn $30 per week in 1929. Speedy work on the hard, chilled meat depended "on the keenness of their knives," which women maintained and sharpened similar to male packinghouse butchers. Their importance to meatpacking firms was such that the U.S. Department of Labor's Women's Bureau could comment during the Great Depression that "the employment office is more reluctant in a slack season to lay off trimmers than any other group of women, because it is difficult to replace them when the business trend is up again." Though not permitted into the masculine killing and cutting departments even when they performed very similar work, skilled female trimmers (almost always older white ethnics) shared job security equivalent to male beef butchers and ham boners.[15]

While the pork trim was a new department, the women who entered sausage manufacturing after 1900 joined a preexisting operation that had been radically altered by technology. In the late nineteenth century a number of equipment supply firms developed meat chopping machinery to speed the sausage manufacturing process. Meat first went through a grinding machine that used a large screw-shaped blade inside a shaft to push meat into a cutting plate that reduced it into chunks about one-eighth inch in diameter. The industrial grinders were enlarged versions of the kitchen devices used by rural housewives, and power-driven rather than hand-operated. Men pushed the meat into the funnel on top and collected it in a wheeled cart when it came out the bottom. These were hazardous machines, as the meat had to be shoveled, and occasionally pushed by hand, into the shaft containing the rotating screw. The Cincinnati Butchers' Supply Company boasted an improved version in which the "oversized hopper" could "be filled without the necessity of tamping the meat into the cylinder, which overcomes the danger of the operator's getting his hands or arms

Sausage chopper,
c. 1900

Ferdinand Ellsworth
Carey, *The Complete
Library of Universal
Knowledge* (1904)

caught in the machine." Such inadvertent recognition of meat grinding's hazards indicates the prevalence of these horrific accidents in early-twentieth-century sausage making.[16]

Once the meat left the grinder, men moved it on small hand trucks to a cutter that reduced it to mincemeat. One early version, Kinyon's "Improved Meat Chopper," combined a rotating wooden base with knives attached to gears that moved up and down. The firm promised that the machine was suitable "for a boy ten years old" and chillingly promised that meat could be taken out and put back in "without stopping, thus *saving much time and labor.*" Large industrial operations in the late nineteenth century generally used an enlarged version using large scythe-shaped blades that chopped the meat in a metal bowl.[17]

By the 1910s the "silent chopper" had displaced these early machines. Versions manufactured by Cincinnati Butchers' Supply and the Allbright-Nell companies shared the same innovation: a rotating metal bowl into which men shoveled the meat, and fitted curved knives that chopped vertically. In the early 1920s Allbright-Nell boasted that its "Buffalo Silent Meat Chopper" could mince 24,000 pounds of meat in a ten-hour working day.[18]

The "finer and more uniform cutting" of the silent choppers may have improved sausage texture, as the Cincinnati Butchers' Supply Company claimed, but also permitted greater manipulation of its contents. A "more liberal use of ice" was possible and necessary with the chopper due to heat generated by the blades as they severed meat's connective tissues. From the manufacturer's perspective, dilution of the meat was not a problem but an opportunity "to produce greater yield and more palatable and juicy sausage."[19] Adding water also permitted including marginal parts of the animal such as snouts, ears, and lips that were otherwise too dry for sausage mixtures. With the added water silent choppers could make the meat weigh more at the end of the process than the beginning!

Fine chopping of meat also invited adulteration. Sausage makers compounded dilution by adding filling agents—usually cereals—that absorbed the water and acted as a binder for the now-tiny meat fragments. Cornmeal could comprise up to 10 percent of hot dogs' weight, and helped conceal high fat content by retarding shrinkage during cooking. Industrial sausages prepared in this way routinely included ingredients to impart a red color as the sausage's interior turned brown once deprived of contact with oxygen. Prior to the 1906 Meat Inspection Act manufacturers mixed in boracic acid, borax, and food coloring and varnished the casings to improve appearance and retard mold growth. The 1906 act banned these ingredients, but manufacturers continued to add saltpeter to the meat and color to the casings to make sausages look fresher. Evidence of gross adulteration through inclusion of spoiled meat or horsemeat is largely allegorical, but the chopping techniques employed in industrial production methods certainly facilitated concealment of these items.

A 1905 "Vienna Frankfurts" recipe indicates how these ingredients ended up in Coney Island red hots. To seventy pounds of shoulder trimmings were added twenty pounds of knuckle meat and sixty pounds from the pig's fat back. This 150-pound mixture (about 40 percent fat) was combined with forty pounds of water and nine pounds of corn flour, along with some salt, sugar, saltpeter, seasonings, and "1 1/2 pounds color water" to impart a red shade to the meat. Another frankfurter recipe called for sixty-five pounds of cheek meat, fifteen pounds of tripe, twenty-five pounds of kidneys, and seventy-seven pounds of regular pork trimmings, with equivalent amounts of water and corn flour. "Always add the corn flour and seasoning and as much water as possible," users of this formula were advised, and do not worry if "the mass would seem very thin" and diluted because the pork trimmings would absorb excess liquid.[20]

Once the "silent chopper" finished mincing the meat, hand labor was again necessary as men pushed carts with the mixture to refrigerated coolers to allow sufficient time for the curing agents to take effect. From the coolers, men loaded the meat by hand into stuffing machines.[21] Holding up to 500 pounds of meat in a tall vertical cylinder, the machines employed an air-actuated piston to push meat through cone-shaped stuffing horns and into the casings. As with all machine-operating jobs associated with sausage manufacturing, this was men's work. In 1913 Cincinnati Butchers' Supply tried to promote its "Boss Air

Sausage line, c. 1900. The men are filling casings with a meat mixture pushed out of the containers at extreme left. Women are twisting the casings into links. Note the supervisors in bowler hats.

Cincinnati Boss Company Records. Courtesy of Archives Center, National Museum of American History, Behring Center, Smithsonian Institution

Stuffer" as a "Great Time Saver. Girls Operate It." Twenty years later they were still stressing that the easy-opening lid permitted a "girl" to run the stuffer. Yet a comprehensive 1932 survey of women meatpacking workers concluded that despite the presence of a few women in these tasks, "the stuffing of sausages is normally a man's job."[22]

Women's work began where the male-operated machinery ended, with linking—separating the casings into the sections that later would become individual sausages. Standing in groups of two, four, six, or eight at tables headed by a stuffing machine and a male operator, women "linked" sausages by applying a "deft twisting motion" to the meat-filled intestines so as to separate the casings at regular intervals. While requiring no tools, it took women several months to become proficient. An adept linker, Upton Sinclair wrote, "worked so fast that the eye could literally not follow her, and there was only a mist of motion, and tangle after tangle of sausages appearing." Until well after World War II, equipment manufacturers were unable to develop automatic linking machinery that could compare with the dexterity, careful touch, and sheer speed of these women. Less skilled women strung the linked sausages onto racks to convey to the smokehouse. Once smoked, more women packed the sausages for shipment.

Linking jobs were good work for women, paying on average $18.47 weekly in 1929 and performed in clean, well-ventilated rooms usually maintained between 40° and 50° F. These positions were dominated by white ethnic women interested in stable employment who, as Sinclair noted, were "apt to have a family to keep alive." The packing and linking jobs, while less well paid, were desirable steps on the job ladder to linking positions. In the late 1920s, two-thirds of all sausage manufacturing workers were or had been married, and the same proportion were over twenty-five. The job had distinct drawbacks, however. Many "linkers" suffered from what we know today as carpal tunnel syndrome because of the feverish pace induced by piecework. And they generally could not move a few feet along the table to operate the mechanical stuffing machines—for which men received ten dollars more per week than the best paid linker.[23]

To create sausage containers, nineteenth-century firms initially purchased casings from companies specializing in "offal," the nonmeat components of the animal. In pre–Civil War New York City the offal was part of a lively secondary meat trade. As part of the fee for using a facility, butchers such as Thomas De Voe provided slaughterhouse owners with offal, which the owners then sold to small businesses specializing in cleaning and preparing the innards for resale. Similarly Cincinnati's pork plants and early Chicago beef slaughterhouses did not handle these parts themselves. They generally sold the offal to small firms specializing in sausage manufacturing or simply discarded the organs. The minimal processing and packaging of the innards in preparation for sale was a job performed by men on one side of the killing floors.

As firms expanded use of offal and intestines, operations became far more detailed and entailed construction of new full-fledged departments, usually located directly underneath the killing floors. Women hired in these areas sorted, cleaned, and routed the animal organs to other areas of the plant, whereas men previously had simply packed them for immediate sale or disposal. The dominant meatpacking union, the Amalgamated Meat Cutters, used the rationale that the jobs were "unfit for any women to perform" to object to female substitution around 1900, but it was too weak to prevent the practice.[24] By the 1930s this department, once considered "unfit" for women workers, could be described by author Tillie Olsen as an area where "men will not work."[25]

"It was terrible," recalled African American packinghouse worker Virginia Houston about work in this department. "That was a dirty job!" Women received the offal from men in the killing areas one floor above through a chute that connected the two departments. The women separated fat that would later be refined into lard or oleomargarine, sent glands to the pharmaceutical area, prepared hearts to be added to sausage filler, and dispatched intestines to be further processed next door in the adjacent casings room. Sixty percent of the offal department's workers were black; white women predominated in these positions only in regions where black women were absent from meatpacking.[26]

The casings room industrialized the operations that Catherine Beecher had advised farm women to use candlesticks to perform. There, women cleaned, salted, and packed intestines for future use as sausage casings. This was a tedious and onerous process that resisted mechanization because of the inconsistent sizes and delicate structure of animal intestines. Similar to the carcass, intestines came in different widths and lengths. Even when cleaning machines came into general use, the eyes and hands of the women workers were essential to make them work properly.

Preparing intestines for use as sausage casings entailed cleaning fat and the "serous membrane" from the exterior and eliminating the contents and mucous lining from the interior while not damaging the double layer of muscle that constituted the intestine wall. Cleaning the outside and removing undigested matter from inside were not difficult operations, though quite foul to perform. Under constant running water, women used their hands to feed a casing stripper that pressed the intestine between rubber rollers, forcing out refuse from inside and stripping off the exterior fat. Women then slid the intestine onto a rod with holes that flushed the interior. "Then you just do that until you get the whole thing on that rod and get it all clean," recalled Houston.

Removing the interior mucous lining was another matter. Intended by the body to ease passage of food to the stomach, the mucus adhered strongly to the intestine's interior. To avoid the slow home technique of turning them inside out, meatpacking plants soaked the intestines overnight to stimulate fermentation of the mucus and muscular casing wall, generating perhaps the worst odors in the entire operation. "The bacterial putrefaction process has always been objectionable because of the irrepressible disagreeable odors" admitted an Allbright-Nell catalog. Tillie Olsen less delicately described the pervasive smell as "the excrement reek of offal" that workers endured "by breathing with open mouth." The process also was risky, as leaving the casings in the solution too long weakened them irreparably.[27]

Once the fermenting casings were removed from the solution, women pushed them through stripping machines two times (interspersed with more soaking in water) before inserting them into a finishing machine designed to purge the interior mucous lining. To complete this process efficiently, the Allbright-Nell Company advised the operator to use her right hand to insert the intestine into the first set of rollers (one corrugated rubber and the other rust-resistant steel) while using her left hand to hold the other end "a foot lower." Once the front of the casing passed through the first set of rollers, the worker directed it between the second set, maintaining enough slack so that it would not break. Water from four spray pipes saturated the rollers (and the employees' arms and hands) throughout the process. Little wonder that Virginia Houston called this a "dirty job!"[28]

Sheep intestines, the preferred frankfurter casings, were especially delicate and labor intensive. In high demand "because the American consumer prefers sausages of a diameter of from ¹¹⁄₁₆ inch to 1 inch," they were very valuable and commanded a "high price." Cleaning was exclusively a hand operation in order to minimize damage and avoid puncturing the intestine wall with "such foreign matter as cinders, glass, or any other ingesta." As a frankfurter-stuffing run's duration was proportional to the sheep casing's length, it is not surprising that supervisors were advised to exercise "constant vigilance to produce the greatest yield, and to preserve at the same time the most attractive color."[29]

Once cleaned and graded (in accordance with width and length), intestines were packed in salt to cure until fully dry, removed, and rubbed with fine salt; only then were they ready to use as sausage containers. Young African American and ethnic women predominated in this department, as older ethnics and native-born with greater job options fled the dirty work for more desirable positions as soon as they could.

The hot dogs of the 1930s were prepared through this arduous, labor-intensive process. The application of machinery did not eliminate close to a dozen stages of physically moving elements of the sausage components from one stage to the next. Hand labor remained an integral part of the process, from the preparation of casings to refining the meat to filling the hot dogs. A complex processing and recombination of animal matter to make frankfurters depended on a highly discontinuous labor process.

Hot dogs also may have "ruled the Ocean side" by the 1930s, but they did not yet have the same popularity at home. Retail vendors, rather than manufacturers, created the principal brands, and most hot dogs purchased in butcher shops were anonymous. They were too insignificant an item to be a census category in the 1930s or in the food consumption surveys of that decade. A 1932 Pittsburgh study found prepackaged sausages for sale only in an upper-class neighborhood, while bulk fresh pork sausage was pervasive in ethnic, African American, and native-born white working-class districts. Ten years later, surveys of eating habits still did not find frankfurter consumption in the home significant enough to chart, collapsing it into the "Bologna, other" category that also included tripe, tongue, and kidney. Through the late 1930s hot dogs' primary appeal was as a food eaten away from home.[30]

Hints of change, though, were in the air. In 1929 Oscar Mayer was the first meat company to brand its franks, using a yellow ring so they "could be distinguished from other wieners in the retailer's case." Its famed Wienermobile, launched in 1936, sought to further the strategy of creating a "personality" for Oscar Mayer wieners. But as a lightly cured cooked product, hot dogs spoiled quickly if not kept cold, and thus shared the same fate as other fresh meats, which consumers preferred to buy as needed from local butcher shops rather than stock in their ice boxes.[31]

As Oscar Mayer's initiatives suggest, packing firms believed that hot dogs had great potential for increased consumption, especially in the home. Swift's choice of the frankfurter to highlight at its World's Fair exhibit is further evidence of manufacturers' hopes for this product. Previously firms had emphasized their branded bacon and hams. But something needed to be done to smooth the production process and persuade consumers they should cook hot dogs themselves. Hence it is not surprising that the industry's trade journal gushed over Swift's World's Fair exhibit.[32]

From the World's Fair to the Frank-O-Matic

For hot dog manufacturers frustrated with the start-and-stop hot-dog-making process and eager to expand the home market for this product, Swift's 1939 World's Fair demonstration was a turning point. The major technological breakthrough demonstrated at the exhibit was in linking previously discrete production stages with conveyer belts and then emulsifying the meat sufficiently to make it a semiliquid batter that could be moved without hand labor at all.

The exhibit was carefully planned both to be a demonstration of production methods for the meats of tomorrow and to promote home hot dog consumption. In his address at the exhibit's opening ceremony, company chairman Charles H. Swift bemoaned the "air of mystery" about sausage, as "people made all sorts of guesses as to what went into them." He promised the exhibit would end "that last lingering bit of mystification" and show the public "the manufacture of one of its universal food favorites."

The exhibit's walls and equipment were white enamel and stainless steel, quite unlike the usual production equipment and surroundings. The women staffing production lines were far younger and more native-born in appearance than most sausage manufacturing workers, and they were dressed in white "well-fitting uniforms," required to wear little white hats and hair nets, and forbidden to chew gum or have "enameled finger nails." Kept out of sight as well was preparation of hot dog casings (they were brought ready to use into the exhibit), hand-linking of sausage, and the water and curing agents added to the meat in the silent cutter. Nor for that matter did the public actually learn what went into the hot dogs. Despite Swift's grand phrases, a great deal of the manufacturing process remained a mystery to curious

Automatic unloading meat-cutting machine

Cincinnati Butchers' Supply Co. *Catalog no. 40* (1946). Cincinnati Boss Company Records. Courtesy of Archives Center, National Museum of American History, Behring Center, Smithsonian Institution

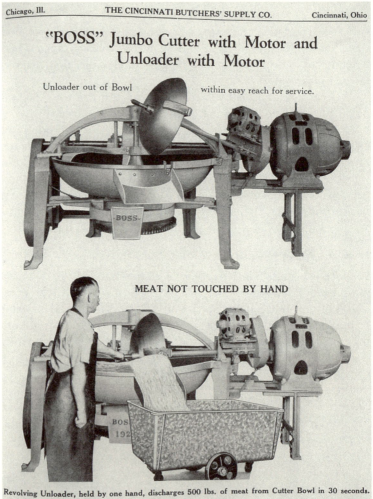

Chicago, Ill. THE CINCINNATI BUTCHERS' SUPPLY CO. Cincinnati, Ohio

"BOSS" Jumbo Cutter with Motor and Unloader with Motor

Unloader out of Bowl   within easy reach for service.

MEAT NOT TOUCHED BY HAND

Revolving Unloader, held by one hand, discharges 500 lbs. of meat from Cutter Bowl in 30 seconds.

Capacity	Floor Space	Horse Power	Weight	Cubic Feet	Code
500 lbs.	6¾x10 ft.	40-50	5500 lbs.	250	Yamig

Motor specified is for Alternating Current, 220 volts, 3 phase, 60 cycles. For motor of other specifications, add code of motor wanted (see page 3).

onlookers—even if they weren't aware of what remained hidden from them!

What the public could see, nonetheless, was a remarkable achievement. The *National Provisioner* praised the integrated production line of Cincinnati Butchers' Supply Company equipment as "a new high in efficient handling." A conveyer belt brought the meat mixture from the cooler directly into the grinder's hopper, and a second belt carried the chopped meat directly to the silent cutter. These belts were an utterly novel innovation for sausage manufacturing, replacing the roustabouts who had previously unloaded and reloaded the meat into these machines. The cutter in turn was high enough so that the "emulsion" flowed directly into the stuffing machine. The *National Provisioner* marveled that "none of the meat is touched by hands" until it is stuffed, thereby reducing the "number of workers required . . . to a minimum."

Building on existing equipment, Swift had made a conceptual breakthrough—integrating the hot dog production line with conveyer belts and emulsifying the sausage meat. Once meat could be reduced to a batter, in essence changed from a solid into a liquid, the opportunities for technological advancement were immense. By the late 1940s Cincinnati Butchers' proudly marketed a series of machines influenced by its World's Fair design. Its new "No. 521" Boss Silent Cutter had "met with the instant approval of the larger sausage makers" because of its speed and efficiency. A conical-shaped unloader could be raised or lowered "with one finger" and took only thirty seconds to empty its contents through a discharge chute. The meat could go directly into a new device, a Boss Lift Dump Mixer that mixed the meat with curing agents. The hopper could then rise three to four feet to a height of eight feet, sufficient to turn over and dump its contents into the sausage stuffing machine. Cincinnati Butchers' promised potential customers that the "saving in labor" would soon offset the cost of this new equipment.[33]

Integrating hot dog manufacturing from chopping to stuffing was a great step forward, but it still left the labor-intensive casing preparation and sausage linking stages relatively unchanged. Speedier meat preparation methods chafed against the restrictions of hand labor at the nether ends of the production lines. Artificial casing manufacturers promised release from these restrictions.

The Visking Corporation had developed artificial hot dog casings fifteen years before Swift's World's Fair exhibit. The demands on this material were substantial: it had to be strong enough to stand the strain of power stuffing, yet easily penetrated by smoke; stable in the presence of water, meat juices, and curing agents, yet sufficiently elastic to expand and contract with its contents during the curing process. Visking

also developed inks to color and label the casings, encouraging brand name promotion of products. In 1950 the company claimed that the availability of this material had meant "Mrs. Housewife's buying ways were quickly changed" as the cellophane casings had "'glamorized' sausage."[34]

From a production standpoint, artificial casings had immense advantages over natural ones. Making them took a fraction of the labor entailed in natural casing preparations, and they could come in immensely long lengths, permitting extended production runs. Producers like Sylvania promised that they would provide "greater uniformity, greater strength, improved appearance." Consumers, however, balked at the obvious disadvantage of this product: it had to be removed from the hot dog before eating, quite unlike edible natural casings. This simple annoyance in the home was a major obstacle for sales in the sites "ruled" by hot dogs because hot dog stands and their operators were not equipped to peel the franks they prepared. Firms using artificial casings thus had to employ workers to remove them by hand, largely obviating increased speed and productivity made possible by this material. With this limitation, it is not surprising that Swift chose not to use cellophane casings in its "hot dog of the future" exhibit.[35]

Automatic removal of artificial casings in the production process was not easy, as it had to be done after hot dogs had been twisted in separate segments, cured, and cooked. A promising machine developed following the war "worked like a charm" at the inventor's plant but failed in a display at a meeting of the American Meat Institute (the principal meatpacking trade association). Equipment put into use late in the 1950s worked better, but due to the "sound and moisture levels" these peelers were housed in separate buildings, once again interfering with a truly continuous production line.[36]

The lure of artificial casings' potential, though, encouraged firms to invest in solutions to the removal problem. The immense benefits of standardizing this meat product were evident: mechanization of previously discrete and labor-intensive stages in production. The process of integrating hot dog production, begun with Swift's 1939 World's Fair exhibit, could be completed.

In the early 1960s, firms solved this technological problem and created hot dog processing operations that needed virtually no workers. The "Frank-O-Matic" took the emulsified meat in one end and fed out perfectly ready hot dogs from the other. Stuffing machines injected the meat mixture into artificial casings carried six abreast into the "processing cabinet" that smoked and cooked them. Automatic pincers, which squeezed the casings to create regularly shaped hot dogs five inches long, replaced the female linkers. Separate compartments for smoking and cooking within the Frank-O-Matic cabinet obviated the need for workers to handle or move the frankfurters from one process to the next. The linked hot dogs—still attached to each other—then were mechanically pulled into a narrow tube-shaped peeler that stripped off the artificial casing—thereby separating the links—and ejected them onto a conveyer belt. The belt conveyed the hot dogs to automatic collating equipment that grouped them according to desired counts prior to sealing in vacuum packaging and labeling for distribution.

The 1960 Frank-O-Matic took less than thirty minutes to size, smoke, cook, chill, and peel a hot dog, using only two operators per machine, and producing almost ten thousand franks per hour. By 1970 Oscar Mayer had a proprietary version of this system that more than tripled this hourly tempo. Its vacuum-sealed "Saran" hot dog package was, company president Oscar Mayer Jr. claimed, "one of the best defenses yet" against meat's natural deterioration. And the annoying jingle developed to sell these wieners seeped into popular consciousness as deeply as hot dogs became a regular part of our diet.[37]

Along with the development of true continuous flow hot dog production came dramatic growth in home consumption. Hot dogs retained their popularity at baseball games and along the beach (though having to compete with new fast foods like hamburgers), but they augmented their reach by becoming a food regularly prepared at home. Hot dogs grew in popularity because they were precooked and easy to prepare. Attractively wrapped in cellophane and ready to eat in just minutes, hot dogs fit nicely as a supplemental lunch or dinner food kept ready for use in the home refrigerators of the 1950s, especially among the baby boomer families with young children. Hot dogs also became standard fare at the barbecue, a widely popular form of private celebratory eating that spread through new suburban developments. In this sense the backyard, as a site for play and consumption, contributed to the steadily increasing popularity of the hot dog, especially for families with children. A 1955 study found that only 13 percent of urban single person households consumed hot dogs at home, while 41 percent of families with children ate them regularly. This was a remarkable expansion of hot dog consumption patterns since the 1930s.[38]

Company advertising reflected an appreciation of hot dogs' new-found popularity for the home market, and an astute awareness of the sites of their greatest popularity. "Summer's a picnic," declared a 1957 Armour newspaper advertisement featuring a cartoon of an idyllic family of four feasting on hot dogs. But lest the consumers think that hot dogs were only for summer occasions, they were assured that there was no need "to wait for a cookout or picnic" to enjoy Armour Star Franks, as the "Open-Fire Flavor is in there for you to enjoy, even if you only heat these franks in water." The advertising campaign's refrain, "Nature makes the meat, Armour makes the difference," sought to brand the franks with quality assurances. "When you can't ask your butcher, you can count on Armour," the advertisements promised, to always deliver

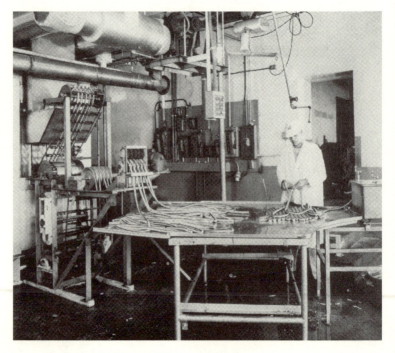

Future hot dogs enter the Frank-O-Matic.

This is the Frank-O-Matic catalog. Courtesy of Smithsonian Institution Libraries, Washington, DC

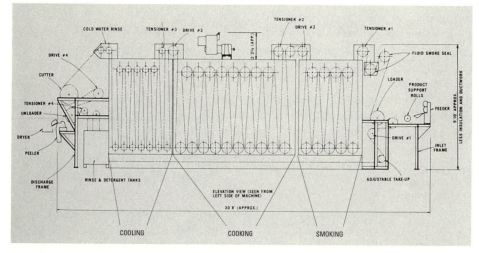

ELEVATION VIEW (SEEN FROM LEFT SIDE OF MACHINE)

COOLING   COOKING   SMOKING

franks "with built-in picknick-y goodness." This meat, while "fun-time serving at its best" also was good to eat, consistently "rich in protein, plump and meaty." While 1930 meat company promotions featured ham and bacon, hot dogs assumed the leading position in company advertising of the late 1950s and early 1960s and were used to help sell other products, such as bacon, canned pork and beans, and breakfast sausage.[39]

Appreciating the influence of children on family hot dog consumption, companies targeted some advertising directly to them during the slow-selling fall and winter months: "When you're knee-high to a hot dog stand, the words 'Armour Star Franks' are the yummiest in the English language." Boys were enticed by an exploding battleship, submarine, and torpedoes that could be obtained for just 35 cents and the backing from an Armour Franks container if sent in before October 31, 1960. An offer oriented to girls expiring March 4, 1961, consisted of a cardboard "Dream House of Dolls" with six individual rooms and figurines depicting stereotyped female scenes in the home. Parents reluctant to be swayed by these promotions were assured that the hot dogs also were good for the children to eat, as they were "made from lean meat, so they're protein rich."[40]

Census data and consumption surveys provide some sense of hot dogs' rapid increase in popularity. The 1963 census reported production of 1.11 billion pounds of "frankfurters and wieners," 30 percent of all sausage made that year. All other sausage varieties were relegated to composite categories such as "dry and semidry" and "other sausage, smoked or cooked," which included bologna. These categories had been recast in the three decades since Swift's exhibit; the 1939 census did not include (or even identify) frankfurters as a meat variety worth

In the Frank-O-Matic, long strings of raw hot dogs in artificial casings enter on the right side and move up and down along rollers through the machine. They are smoked in the set of four rollers at right, then move into the larger central chamber for cooking before cooling in the last section. When they exit the machine at left the cutter automatically separates the links and the peeler removes the artificial casings.

This is the Frank-O-Matic catalog. Courtesy of Smithsonian Institution Libraries, Washington, DC

recording.[41] A similar conceptual change took place in consumption studies. While mid-1930s surveys did not separate out frankfurters from the general "sausage" category, thirty years later frankfurters were the only variety of lunch meat and sausage to be identified by name. A 1965 study by the U.S. Department of Agriculture showed that household per capita hot dog consumption averaged nine pounds, or about seventy-five hot dogs, per year, and held remarkably consistent across region and socioeconomic group. Clearly what had once been one cured product among many had emerged as America's preeminent sausage.

Hot dogs' dizzying rise to an everyday food item after World War II was no secret. It was evident in advertising, the branded varieties from manufactures available in supermarkets, and the consistent place of hot dogs in the family dining cycle. A far less evident consequence was workplace turmoil generated by the transformation in the production process. Dramatic automation of hot dog production eliminated the jobs of thousands of women. Automatic machines requiring just two operators replaced the linkers, hangers, and packers who had once filled sausage departments. "Oh yes, they was getting rid of the women," recalled Velma Otterman Schrader. In her Waterloo, Iowa, Rath plant the new machinery reduced a department formerly comprising three hundred women to just twelve. Pictures of sausage departments in the 1970s show only men; the women who had labored in this area for half a century had all but vanished.[42]

Job losses were less dramatic in pork trim because the meat continued to come in irregular sizes best handled by an experienced eye and practiced hand. But even there the expanded use of mechanical "whizzard" knives with a moving circular blade reduced skill requirements and made this a job that could be learned in a few weeks. Casings operations were not so much mechanized as eliminated, as firms no longer need these products to package their sausages.

The changes caused massive job losses for female packinghouse workers in the early 1960s at the same time as hot dog consumption soared. Because company policies and collective bargaining agreements did not permit women to move into jobs traditionally held by men, displaced women workers who had made sausages for many years watched in fury as men with far less service remained at work in the same plant. Women who had worked together for decades were laid off together; quite quickly they banded together, seeking to regain jobs in their plant, if necessary in areas when men traditionally worked. When the union and company proved stubbornly unresponsive, the women turned to the 1965 Civil Rights Act and filed suits under Title VII alleging sex discrimination in employment. Hundreds of female packinghouse workers provided supporting complaints to the Equal Employment Opportunity Commission (EEOC) that interpreted and enforced the Act.

The complaints from female packinghouse workers had a dramatic impact on the EEOC and the interpretation of the Civil Rights Act. While the Act did not explicitly prohibit limiting job opportunities by sex, the hundreds of cases presented to the EEOC from displaced sausage manufacturing workers persuaded the agency that this practice was inherently discriminatory. The EEOC ruled in 1969 that Title VII banned designation of jobs by sex throughout the U.S. economy. By 1970 most displaced female packinghouse workers were able to regain employment, albeit in different, often less desirable, jobs. Their protests, though, had an impact far beyond their industry, as it no longer was legal for firms or labor organizations to designate some jobs for men and some for women. Inadvertently, yet dramatically, efforts by firms to mass produce hot dogs had led to a struggle against sex discrimination, and a major advance for women's employment rights in America.

The same year that the EEOC found female sausage workers had been treated unfairly, the lowly hot dog became a target of public debate at the highest levels of American government. At the 1969 hearings of the Senate Committee on Nutrition and Human Needs, Ralph Nader and other consumer advocates made the frankfurter an archetype symbol for the degradation of America's meat supply. Calling it the "fatfurter," Nader described the hot dog as a form of food designed to "defraud" the consumer by selling her "substandard meat palatably." Nader especially targeted the fat content of hot dogs, alleging that some contained as much as 40 percent fat. No longer could the hot dog simply serve as a "fun" food; its very popularity now meant it had to pass higher scrutiny for its wholesomeness and nutritional content.[43]

Not all hot dogs were guilty of Nader's accusations, but enough were that the government instituted controls over fat content and other ingredients. The source of the problem was the very same production methods that had vaulted the hot dog to such prominence among American meats. Sausage manufacturers still followed the adage that they should "interchange" the meat varieties used in their products "as the economics change" so they could make a healthy profit. Yet they also had an even greater imperative to make a "uniform product" that tasted the same to consumers and could be processed in the automatic production systems. These were competing objectives, quite at odds, and they pushed the composition of hot dogs in the direction decried by Nader.[44]

Emulsifying meat of constantly varying combinations reflected the perennial dilemma of meat producers, subduing the natural variations in the material they had to work with. The silent cutter was the key step in this process, as it could shred the meat until it "has the consistency of bread dough." Adding hard dry meat, such as from bulls or the face meat of cows, required balancing with large quantities of fatty meat (usually pork) so that the mixture would have the proper texture. Temperature also was an obstacle; cutting the meat so finely generated considerable heat, requiring addition of ice during the cutting process. Too little ice and the mixture would exceed temperatures permitting proper emulsification; too much ice and it would not hold together. To help bind the mixture sufficiently so that it "stuffs tighter" and "sets-up firmer," companies added flour or cornmeal, especially if the meat was of inferior quality. Miscalculating this process could ruin the mixture, turning it into "solid lumps" if not cut enough, or "runny" if it was too soft.[45]

Production equipment such as the Frank-O-Matic increased pressures on meat firms to combine "perfect emulsification" with "complete retention of moisture" in order to "yield a product of uniform juiciness, tenderness and good flavor." A poor batch was a large problem for the older sausage makers, but for firms producing 30,000 per hour it was a disaster. One solution, standardizing the types of meat used for hot dogs, clashed with the principal value of sausage products, their capacity to recreate and recombine meats from many different sources. More often manufacturers gave themselves a greater cushion by increasing the moisture and fat content so as to accommodate the variations in meat input. Adding binding material accompanied this production decision, as flours and grains minimized shrinkage during cooking of these moisture- and fat-laden hot dogs.[46]

The adverse publicity generated by Nader's appearance, and the accompanying flurry of books and articles from "Nader's Raiders," frustrated meat producers. As experience had taught them that "the tenderness and juiciness of frankfurters and bologna are directly related to the content of both fat and moisture," many felt they were being pilloried for doing no more than giving the customer what she wanted. Skyrocketing production numbers and the popularity of their hot dogs from baseball games to backyard barbecues seemed at odds with their portrayal in the media and before Congress. They did not like to see themselves as using hot dogs to "sell substandard meat palatably," as Nader charged. Yet both perspectives were true at the same time—the hot dog was immensely popular and meat producers were using hot dogs in the way meat producers had always used sausages, to sell "low priced" animal parts not as "palatable" as "porterhouse steak" and placing them in a recreated package that concealed the origins of its contents.[47]

The signature hot dog advertising campaign, Oscar Mayer's "I wish I was an Oscar Mayer wiener" ditty, can be loosely interpreted as an expression of meat company's fantasies that more types of meat could be like the hot dog. Unlike the beef and pork cuts that continued to come in irregular sizes and thereby pose major impediments to mechanizing production, the meat in hot dogs could be emulsified, permitting creation of a standard sized meat product. Once true standardization could be attained, mechanization of production could occur to a degree previously inconceivable in the meat industry.

It was a long road to the standardized frankfurter. The first stage was applying machinery to the discrete processes farmwomen had previously performed by hand. Its product distinguished by commingling meat from different animals, turn-of-the-century sausage manufacturing permitted a level of processing technology far exceeding what was used with other meats. New machinery and fine division of labor stimulated vast increases in productivity, and hot dogs became a fixture at places of public leisure by the 1930s.

To take the next step, processors had to move beyond mechanizing discrete production stages and imagine integrating those stages of production. Swift broke through this barrier at the 1939 World's Fair, and postwar innovations in materials and technology allowed true continuous process production. By the mid-1960s hot dogs had improved on their 1930s popularity to spread into the homes of families, especially those with children, and to assume a favored place in their lunch and dinner meal cycle.

Revolutionizing hot dog production had profound unintentional consequences. The layoffs of female sausage makers in the 1960s stimulated working women's protests that contributed to the 1969 EEOC decision prohibiting sex-based job restrictions throughout the American economy. And the production changes stimulated alterations in the composition of hot dogs that brought its nutritional value into question and encouraged close regulation of its contents by the federal government. Yet consumption remains steady; contemporary estimates place hot dog consumption at 50 million per day, for an average of eighty per person per year.[48]

The contents of hot dogs remain mysterious to most consumers, even after federal labeling requirements mandated that packages indicate the source of the meat they contain. While processors eschew the more disturbing components common to turn-of-the-century frankfurters (such as lungs), the "art of sausage making" remains now, as it was in 1905, taking otherwise unusable parts of the cattle and pig carcass "and making from them a palatable, wholesome and at the same time economical article." Encapsulating both ancient traditions and modern industrial technology, hot dogs were another meat transformed by the interface of company production innovations and changing consumer demand. They were not a new meat, however; to that honor must go the chicken.

Chapter Five

Chicken

In June 1948 an enthusiastic three-mile parade wended its way through the tiny town of Georgetown, Delaware, as the final event in the improbably named (to contemporary ears) "Del-Mar-Va Chicken of Tomorrow Festival." The parade celebrated a remarkable event that had been building for several years—the national "Chicken of Tomorrow" contest. A national committee of poultry industry organizations promoted the contest, which was initiated by the A&P retail chain and the U.S. Department of Agriculture, to encourage "production of superior meat-type chickens." A series of state and regional contests provided cash prizes to winners and determined qualified entries for the national competition in Georgetown.[1]

Forty entrants from leading hatcheries throughout the United States competed for the lucrative national prize. Not only would the winner receive $5,000 but doubtless also many orders from farmers eager to grow the best birds for the market. The winner, the Vantress Hatchery in California, was able to grow a heavier, meatier chicken faster than any other entrant. Within ten years the Vantress-produced birds would be the standard used by the nation's poultry farmers.[2]

The post–World War II era saw a fundamental transition in the place of chicken within America's diet, and, indeed, in the popular conception of chicken as a type of food. For two hundred years Americans considered chicken a luxury meat served only on special occasions. The Republican Party's unfortunate 1928 campaign slogan, "A Chicken in Every Pot and a Car in Every Garage," reflected chicken's status as an unusual, expensive, and hard-to-obtain food. Americans ate only fifteen pounds per year when Herbert Hoover campaigned with the promise to make chicken more widely available, and they paid thirty-eight cents per pound for the bird, about the same price as round steak and more expensive than fresh pork chops or ham.[3]

By the 1980s chicken was challenging beef as America's favorite meat. Rarely has there been such a dramatic change in American foodways. Vastly increased consumption practices took place in several arenas. The fast food industry, led by Kentucky Fried Chicken, established chicken as a meal to be eaten quickly and for lunch. Simultaneously, concerns over red meat's fat content, an outgrowth of the health and consumers' movement, resulted in consumers shifting their eating preferences from beef to chicken. (It is, of course, ironic that two contradictory trends in American foodways redounded to the benefit of the poultry industry.) What had once been a food item eaten on Sunday was now the center of several meals each week.

The very language changed along with these consumption habits. At least through the 1940s chicken was part of a larger category called poultry or fowl that contained many distinct breeds. Most farmers relied on White Leghorns for egg production, but many other varieties circulated through the nation's farms and meat markets. Broilers referred not to a type of chicken, but to a stage of development that suited the animal to a certain kind of cooking. Aggressive cross-breeding of chicken following the war largely eliminated breeds outside of poultry fanciers, in favor of distinctions by form in which the animal would be used: layers, broilers, roasters, and so forth.

Paralleling the metamorphosis of chicken as an animal category distinguished by function rather than lineage was the incorporation of chicken into the category of meat. Poultry firms helped transform chicken's place in America's diet by literally changing the form in which consumers encountered it in eating establishments and supermarkets. This entailed a conceptual shift that the physical integrity of the chicken could be violated to create new products; that the "meat-type chicken" could be transformed into many kinds of chicken meat.

From Eggs to Broilers

Poultry consumption has a long history as an annex to the rural cycle of harvesting eggs for home use and the market. Chickens are relatively easy to raise and keep, and for two centuries both rural and urban areas had plenty on hand. Data is especially unreliable for chickens, but they were ubiquitous in nineteenth-century America. The U.S. census of 1840 estimated the value of poultry on farms at $12 million and their numbers and popularity rose steadily throughout the century. In 1910, 88 percent of all farmers kept chickens, with an average flock of around eighty.[4]

Eggs, rather than chickens, were the preferred commercial products, for the evident value of offering a constantly replenishing supply of high-quality protein. Availability was highly seasonal until the twentieth century, as egg production was strongest in the spring and summer and tapered off through the rest of the year. Nineteenth-century household advice books contained various suggestions on how to keep eggs fresh through the year, such as packing them in barley or keeping them in a cool place. By the early twentieth century the large meatpacking firms entered the egg business aggressively. Their chilled railroad cars and national branch house network were well suited to taking the seasonal products of the farm and making them available year round for urban consumers.

Early-nineteenth-century chicken meat generally came in two forms. In the late spring and summer, farmers would cull the excess cockerels
(young males) when, Miss Leslie advised, their "body should be thick and the breast fat," eat some for home consumption, and sell the rest. Generally weighing 1½ to 3 pounds, these "young and fine" chickens with their "thin and tender" skin were the best birds to eat. While cockerels (forerunners of today's broilers, fryers, and roasters) were a seasonal specialty, laying hens could be harvested year round, usually after their egg-producing capacity declined. Their meat was less desirable, as "Old poultry is tough and hard." While young birds could be roasted or fried, the hens needed to be boiled for hours to ensure the meat was palatable.[5]

Recipes reflect preparation methods that shielded consumers against the uncertainties of chicken meat quality. Boiling and stewing predominated in pre–Civil War manuals, as use of wet heat allowed for more predictable slow-cooking methods suited to softening tougher flesh. Miss Leslie's recipes, clearly anticipating use of an open hearth, suggested a system where the fowl sat in water inside a larger container kept at a hard boil for several hours. Various kinds of heavy gravies accompanied the stewed chicken meat when served.[6] As cooking stoves became commonplace in the latter part of the nineteenth century, broiling and frying recipes became more prevalent, reflecting the greater ease of these preparations on the stove top or in interior compartments. Better control over heat levels was critical to these cooking methods for chicken, as broiling required consistent "hot and bright coals" while frying "can be properly done only when the fat is smoking hot." Uncertainty over chicken quality, though, was reflected in admonitions such as "young spring chickens only are used for broiling" and warnings to use "a tender chicken" for frying. The fricassee was a popular compromise in which the chicken was first boiled to be tenderized and then fried in butter and spices to add flavor, thereby reducing dependence on tender chicken varieties.[7]

Urban demand for poultry in the nineteenth century stimulated farmers to bring fowl and eggs to city markets along with their produce. Doubtless many planned to have sufficient chickens to supply family needs (especially in the summer months when fresh pork and beef were not available), as well as to provide a small surplus for cash income. In 1867 Thomas De Voe observed that Pennsylvania was a major supplier of dressed chicken to New York markets "in the cold season," while during spring and summer many regional farmers live-shipped their birds to market. They were killed on demand for shoppers and sold with the feathers removed but the head, feet, and innards intact. "Dry-picking" produced a better meat than the more common method of first scalding the chicken to loosen the feathers, but it required techniques that even an experienced butcher like De Voe felt were "unnecessarily cruel." Stuck with a pin through the upper jaw into the brain, the struggling animal was kept alive while the feathers were pulled off so that they did not set firmly in the flesh.[8]

Quality was highly uneven, however. Many different breeds at different weights and maturity were on sale at any one time, demanding considerable skill among shoppers to discern the best birds. Commanding the highest prices were desexed males, the "caponed fowl" that combined "the tenderness of the chicken with the fine juicy flavor of maturity." De Voe especially endorsed the Dorkings and Cochin China varieties for this "delicacy." Closely following in price were broilers, also known as spring chickens, generally males who naturally were no use for egg laying. As most were hatched in February, March, and April they were brought to market in the spring and not generally available year round. "There are few species of the bird kind more tender than a young chicken," De Voe observed. Discerning the caponed fowl or broiler from the older chicken was very important, as there were few birds "tougher than an old hen or cock" over a year old.[9]

By the 1880s thousands of farmers in the northeast were specializing in meat chicken production for the urban markets of the Washington-Boston corridor. These operations remained marginal, however, as they could not deliver birds of consistent quality to urban consumers. The best young eating chickens were a seasonal specialty; for much of the year consumers never quite knew what would be available when they went to market. The prevalence of stewed and boiled chicken at the turn of the century reflected the uncertainties of the retail market, the difficulty of finding—at a good price—the kind of tender chickens that could be broiled or fried.[10]

Swift entered this burgeoning market in the early twentieth century with the hope that its national structures would permit greater standardization of the chicken that went to market. The firm produced several different lines but, similar to its branding strategy in the pork industry, aggressively promoted the high-end "Premium Milk-Fed Chicken," as a "delicacy . . . fit for the tables of kings." Collected from farmers at its St. Joseph, Missouri, plant, the chickens were fattened for the last two weeks of their lives on buttermilk and grain. Birds reluctant to eat sufficient quantities were "crammed" with a device resembling "a large foot-pump for a bicycle," according to an appreciative description in the *Ladies' Home Journal.* An operator removed a bird from its cage and inserted a foot-long rubber tube through the bird's mouth into its stomach, regulating the feeding pressure by feeling the bird's crop just "as a bicyclist feels of his tire when pumping it up." Given "the

Poultry plucking,
c. 1900

Ferdinand Ellsworth
Carey, *The Complete
Library of Universal
Knowledge* (1904)

squarest meal he ever had in his life" in just ten seconds, the bird was
fed twice daily until ready for slaughter.[11]

Swift's primitive slaughtering methods belied the *Ladies' Home Journal*'s commentary that its handling of chickens contrasted markedly
with the "questionable care and unsanitary conditions of the barn-
yards where they are merely so much 'pin money' for the farmer's wife."
Workers hung birds by their feet from a moving chain and killed them
with a long, thin knife stuck through their gullet. The chain carried the
carcass past a line of twenty men who each removed a few feathers before the chicken went in a cooler for chilling (and sometimes freezing),
until shipped out by railroad car and distributed through Swift's branch
house system. (Less desirable birds went through a scalding tub that
accelerated the feather removal process.) In the "New York Dressed"
form that dominated poultry production until the 1960s, intestines,
internal organs, the head and feet were not removed. Plants of this type
could handle ten thousand chickens per day around 1900.[12]

Investment in chicken processing reflected Swift's belief that there
was a strong demand for "Premium Poultry" regardless of the cost. The
buttermilk fattening process sought to make the flesh "tender and
juicy" and to "lighten" the dark meat until it was "but little darker than
the 'white meat' of ordinary fowl." Its promotional language and
processing methods sought to engage with early-twentieth-century
assumptions concerning what made for good eating chicken.[13]

While mid-nineteenth-century farmers knew the age and condi-
tion of the chickens they ate, shoppers in search of chicken in twentieth-century butcher shops had a set of visual criteria that paralleled
their concerns for beef's color and odor. Buyers looked for bright eyes,

soft feet, and skin "of a clear color—a yellow tint being best liked." Consumers opted for light colors in chickens, as darker meat was evidence of an older (and potentially tough) bird. Leaving on the feet and head was in part a response to consumers' interest in assessing the bird's quality, as well as exempting the processor from hiring the labor to perform these operations.[14]

National meat processors such as Swift and Wilson were powerful forces in the national market for frozen chicken that somewhat reduced the seasonal character of chicken supplies. But the widespread availability of chickens on farms and corresponding dispersion of slaughtering and dressing among urban centers precluded centralization of the poultry industry paralleling beef and pork. Instead, as demand grew so too did highly regional chicken markets, as adjacent farming regions learned to supply the needs of growing urban centers.

New York City was the largest market for chicken in the nation, in part because it remained America's biggest city, and in part because of its peculiar ethnic composition. Forbidden from eating pork by kosher dietary rules, New York's Jews were eager consumers of poultry in order to add variety to their diets and to have a special meal for Sundays. A 1926 Department of Agriculture study found that Jews accounted for 80 percent of the live poultry sales in New York City; with a Jewish population of two million by the 1930s, this was a substantial market. Yet they could not partake of the Midwestern slaughtered chickens. Orthodox Jews only would eat chicken killed by licensed *shoctim* using the prescribed kosher method of cutting the gullet and windpipe with two quick forward and backward strokes, then piercing the veins on both sides of the neck.[15]

Farmers in the Delmarva peninsula were the principal beneficiaries of the burgeoning demand for chicken (Delmarva refers to the peninsula lying between the Chesapeake and Atlantic that includes Delaware, Maryland's eastern shore, and two Virginia counties). In fact, Jewish demand for live chicken fundamentally changed Delmarva agriculture. Discovery of this market was an accident generally attributed to Cecile Steele, a Sussex County, Delaware, farmwoman who maintained a flock of laying chickens to contribute to her family's income. In 1923 she mistakenly received five hundred chicks from a hatchery, ten times her usual order. She raised the chickens rather than send them back, and eighteen weeks later was able to receive 62 cents a pound for them, a huge profit. Steele invested some of her earnings in an order for a thousand chicks; within three years she and her husband had expanded their capacity to ten thousand. Their success encouraged other struggling farmers to explore this market. By the end of the 1920s there were five hundred chicken growers in Sussex County. While husbands and children may have become involved in this business, women remained

Poultry processing, Sussex County, Delaware, 1940s

Courtesy Delaware State Public Archives

central to raising the broilers for market. In the mid-1930s virtually all of Delaware's chickens, produced by Protestants who had lived for generations in the same area, went to New York City for the Jewish immigrant market.

For the first ten years of this new business most chickens were "live-shipped" to New York commission markets, where brokers bought them in lots and in turn distributed the birds to local stores. Beginning in the late 1930s the first processing plants opened in Delaware, and by 1942 there were ten in operation with the capacity to process 38 million broilers annually. There was an ample Gentile market in New York for these birds, along with less strict second-generation Jews who were willing to patronize kosher butchers who sold fresh-killed chicken shipped in ice.[16]

These plants were primarily hand operations that echoed the methods of late-nineteenth-century hog processing operations. Men first hung the animals by their legs from racks attached to a moving chain that carried them to successive operations. Women cut chickens' throats with a knife, and once the animal was bled it entered the five-stage feather-removal process. Just as firms had figured out how to replace the men who had shaved the hair from a hog's skin, poultry equipment suppliers developed machinery that supplanted the men and women who had pulled out the feathers by hand. The birds were scalded, whirled around in a drum with rubber "fingers," coated in wax and picked over by women to eliminate most of the feathers, then finally singed with flame to clean the last remnants. Female consumers or their butchers performed the final processing stage of evisceration on these "New York Dressed" birds prior to cooking. Plants employing these methods could process a hundred thousand chickens a day in 1945.[17]

More routine availability of chicken entrenched that meat's place in the national diet during the first half of the twentieth century. Refrigeration attenuated nature's influence on meat quality by slowing decomposition, as cold storage allowed frozen chickens (albeit of inferior quality) to be used when fresh supplies waned. Broilers were still "spring chickens"; a 1935 manual for hotels advised that the "fresh broiler season" ran from June until October, while the slightly heavier roasters were available, fresh, from September through January. Refrigeration did mean that there was no true "closed season" for spring chicken, even though the guide admitted that this variety was "more injured by cold storage" than any other class of poultry."[18]

The simple recipes of the early twentieth century were elaborated in cookbooks at mid-century devoted entirely to poultry. Such an approach indicated an important shift in chicken preparation. These recipe books assumed that chicken was available year round and that there was a need to promote new ways of preparing it so that the taste would not be monotonous. James Beard published *Fowl and Game Cookery* in 1942 to promote more supple preparation of chicken as there was "no other item of our food has so many possibilities nor is there anything as quickly and simply prepared." His stated mission was to expand chicken consumption beyond the fried chicken available in roadside venues whose signs along American roads could "supply material for a fleet of boats." Beard suggested extensive use of sauces over lightly fried, sautéed, and broiled chicken, and incorporating ethnic spicing into traditional stews to create Spanish paella and Italian chicken casserole.[19]

Almost ten years later, *The Complete Chicken Cookery* more than doubled the number of recipes in Beard's book with a similar method, using varieties of sauces and ingredients to diversify the taste of chicken cooked through traditional dry and wet heat methods. "The variations are almost literally infinite," promised Marian Tracy, as chicken could satisfy "sophisticates" as well as those with "simple tastes." Promoting far wider forms of chicken consumption so that it would become America's leading meat protein was the principal objective of the 1948 Chicken of Tomorrow contest.[20]

Remaking the Postwar Chicken

The "Chicken of Tomorrow" event brought together major institutions that would, collectively, transform the place of chicken in American cuisine in the last half of the twentieth century. A well-established net-work of chicken breeders, egg hatcheries, feed producers, and process-ing and distributing firms hoped that the postwar years would vastly expand the market for their products. Large retail chains, principally the East Coast A&P firm, wanted a better product for consumers now interested in regular chicken purchases. But the fragmentation of actors, and conflicts among them, meant that it took an institution both embedded in yet not "of" the industry to create a shared strategy—the U.S. Department of Agriculture and its Cooperative Extension Service.

Established by the 1887 Hatch Act, state extension services were attached to the land grant colleges of each state with the charge to assist the farm economy. The network of county agents and poultry special-ists encouraged farmers whose fruit and cotton crops were damaged by bugs and diseases in the 1920s to raise broilers. Disseminating techni-cal information on matters such as proper food mixtures and chicken house design, extension service personnel also acted as mediators be-tween the emerging urban markets and the rural growers.[21] Following the war, state extension services went into high gear to persuade farm-ers they could profit from promised postwar prosperity by improving their chickens' quality and reducing production costs. In the Deep South extension personnel encouraged troubled cotton farmers to shift into broiler production.

Georgetown's "Chicken of Tomorrow" festival and contest were in large part due to the efforts of Delaware's Extension Service and its poul-try specialist, J. Frank Gordy Jr. In their modernizing project Delaware extension personnel stressed the importance of consumer demand to "grower, processor, hatcheryman, feed dealer." Farmers had a "duty to use growing practices that tend to produce better market quality . . . even though it may be to his short-run advantage to lower his stan-dards." Poultry processors had to be constrained against lower-quality production, as it "has a detrimental effect on all prices" and can give Delmarva chickens "an unfavorable reputation." The extension service not only promoted better chicken-producing habits but also sought to make consumer demand a principal factor in the decisions of all indus-try sectors.[22]

Gordy's role indicates the valuable nonpartisan status of the exten-sion service among poultry interests. A central part of the industry yet tied to no particular firm or sector, the extension service could play a unique role advancing commercial broiler production. Prior to World War II, the extension service's most important assistance to chicken producers was in the area of disease control. Because it was part of the state's land grant institution, the University of Delaware, the extension service could call on university faculty to assist its scientific studies. Gordy built on these established relationships and trust to expand extension service resources that could be devoted to assisting the poultry industry.

He was above all concerned with restoring Delaware's flagging place in the national industry. Prior to World War II, Delaware chickens had dominated the important New York City market and were the main poultry product in other East Coast and Midwest cities. During wartime, the armed forces requisitioned Delmarva's entire chicken production for military use, disrupting these relationships and allowing Georgia and Arkansas growers to gain footholds in lucrative urban markets. (Military buying also encouraged a flourishing black market as poultry growers tried to evade price ceilings and get their product to New York City outlets.) Following the war chicken producers in New England and the Deep South rapidly expanded their sales in midwestern and New England states. Although Delmarva's broiler production held steady, its market share fell from 25 percent in 1940 to only 6 percent in 1955 as national broiler production topped one billion for the first time.[23]

One of Gordy's first initiatives was to persuade a new university faculty member, Willard McAllister, to investigate how Delaware chicken was faring in the retail market. Beginning with a painstaking canvass of food retailers in Philadelphia, McAllister's careful studies identified a series of challenges for poultry producers in the late 1940s. He determined that increasing consumption would not take place until there was a change in the popular attitude that chicken was a special weekend meal. Before chicken could become a more regular and consistent part of the American diet, McAllister argued that the price of broilers had to be reduced below that of beef and pork and the quality of its meat improved. This meant producing an inexpensive and "meatier" bird that had a fresh and attractive appearance in the retail store. McAllister's studies also showed that retailers and consumers preferred a bird that was fully processed and ready to cook, rather than the traditional "New York dressed" style.[24]

Implementing McAllister's recommendations to produce a better broiler entailed dramatic alterations in farming and processing methods—and in the bird itself. To persuade the fragmented and quarrelsome poultrymen of his program, Gordy diligently worked to engender a greater consciousness of the trouble they were in and the urgent need for cooperation among different sectors.

For an extension service rooted in imparting technical advice, the efforts of Gordy and his associates were remarkable. Publications oriented to the farmer, such as "Mr. Poultryman: Marketing Is Your Business," and junkets to wholesale poultry markets in New York City and Philadelphia were designed to persuade growers that they had to change their methods to respond to consumer preferences. These far-reaching efforts even extended to using the 4-H program to influence future farmers (and their parents) through a "Junior Broiler Program" contest modeled on the "Chicken of Tomorrow" competition.[25]

Gordy and other extension service personnel avidly participated in the Chicken of Tomorrow contest as a first step in their campaign to

persuade consumers that chicken was an everyday meal. However, widespread acceptance of chicks from the contest winner was another matter. Farmers were reluctant to buy the more expensive Vantress birds and responded sluggishly to other aspects of McAllister's recommendations, such as switching to white birds so that small pinfeathers not removed in the processing plant would be less visible to consumers. To change the chicken, Gordy and his associates also needed leverage to persuade farmers to change their chicken-raising practices.

Establishing a chicken auction in 1952, the Eastern Shore Poultry Growers' Exchange, was an integral part of Gordy's strategy. By creating an open market in broilers, the extension service hoped that market forces would induce farmers to improve their chickens' quality and accede to chicken processors' preferences. Half a million chickens were sold on June 24, its first day of business, to local poultry processing plants; within five years, daily sales would frequently top one million. Very quickly the auction became "part of the social fabric" of life on the Delmarva peninsula. In Selbyville, hundreds of growers, feed dealers, buyers, and other "interested persons" would gather at the exchange building at 1 P.M. to learn the auction results. Those unable to attend would turn on their radios as popular stations broadcast the latest chicken prices with the same drama that television announces today's winning lottery numbers. The Delmarva exchange was quickly emulated in other poultry-growing states such as Arkansas.[26]

The exchange motivated farmers to comply with the extension service's recommendations if they wanted to receive good prices for their flocks. "Selling birds through the auction has emphasized difference in price for birds of different quality far more than the average grower could realize before the auction was formed," Gordy noted approvingly. Chickens produced from Vantress stock (the winners of the Chicken of Tomorrow contest) grew from 12 percent of auction sales in 1953 to 76 percent in 1957, and sales of white-feather chickens increased from 52 percent to over 80 percent in the same period. At the same time, the average age of chickens brought to market fell from twelve to nine weeks, reflecting improved breeding stock and feeding methods. There also were significant reductions in mortality rates and food consumption costs.[27]

These improvements came at a price to farmers—to survive in the poultry business they had to sacrifice their traditional autonomy. Building better chicken houses, buying more expensive chicks, and providing improved feeds all required capital that farmers did not have. In the

1930s and 1940s farmers paid effective annual interest rates of 15 to 25 percent to obtain hatchling chickens and necessary supplies, quite onerous and also insufficient to acquire new farming technology. In the 1950s most farmers in Delaware, and throughout the southern poultry region, switched to contracting arrangements with feed supply companies to finance their operations. Under these contracts feed companies retained ownership of the birds and (through agents known as servicemen) specified feeds and other raising practices employed by the farmer. Your companies "are putting in 80% of the risk capital," Willard McAllister admonished the servicemen, "you should have at least this much control of the growing operation." Although contracting protected farmers from natural disasters and ruinous interest charges, it also deprived them of a great deal of independence, as they had to accept the chicken-raising practices mandated by their suppliers.[28]

By the end of the 1950s, the extension service began to advocate the creation of firms that oversaw broiler production from the egg to the processed carcass—usually termed "integration"—as the best way to improve the Delmarva industry. Its personnel believed that fragmentation of the industry into complex factions of egg hatcheries, feed producers, growers, and processors impeded necessary restructuring of chicken production. Practices "which may make short term profits for an individual or a separate segment of the total industry are minimized in an integrated firm," McAllister explained in "A Plan of Action for the Delmarva Poultry Industry." He concluded on an optimistic note, "Therefore, any wasteful or costly practice is not tolerated."[29]

The majority of the new large chicken firms had their origins on the agricultural side of the industry. Incorporating processing operations usually was the last step to building an integrated company. The formation of the Townsend poultry company is a good illustration of this process. A wealthy agricultural family that sent a member, John G., to the U.S. Senate in the 1930s, the Townsends began financing farmers to raise poultry in the mid-1930s. By the end of World War II the company had its own chicken hatchery and feed mill and had expanded contracts with local farmers to produce chickens, using its chicks and feed. It sold chickens through the Eastern Shore Poultry Growers' Exchange until the late 1950s, when the firm built its own processing plant. Most Delmarva poultry producers soon followed this example, paralleling similar trends among Deep South chicken firms. By the 1960s these new integrated companies would force out older meat companies such as Swift and dominate the invigorated poultry industry.[30]

Ironically, the extension service's success would doom its creation, the Eastern Shore Poultry Grower's Exchange. As integrated poultry growing operations acquired processing plants and internalized broiler production stages, the exchange's sales volume precipitously declined.

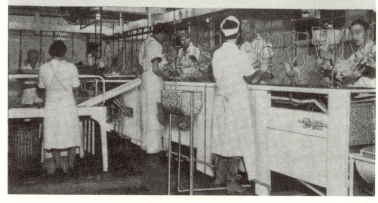

SANITARY POULTRY EVISCERATING EQUIPMENT

Early evisceration line, 1940s

Barker Poultry Equipment Co., *The Barker Catalog no. 44* (1946). Courtesy of Smithsonian Institution Libraries, Washington, DC

The end came when the last major grower to use the exchange, Frank Perdue, finally acquired a Swift processing plant in 1968. He almost immediately stopped selling flocks through the exchange because, as Perdue later recalled, "there was more money in processing." The exchange closed soon thereafter, as it was now the integrated firms that enforced chicken quality.[31]

The emergence of the "integrators," firms like Perdue and Townsend on the Delmarva peninsula and Tyson in Arkansas, led to renewed attention to processing methods. Until the late 1950s most chicken left the processing plant as uneviscerated "New York dressed" poultry. The facilities handling these birds were rudimentary operations with relatively low labor demands, able to rely largely on female workers from nearby rural areas. The expansion of processing to include evisceration and better cleaning of the carcasses, stimulated by the advent of federal inspection in 1959 and the growing use of chain supermarkets to sell chickens, resulted in construction of brand-new facilities.

Evisceration also added enormously to the industry's labor needs. So-called on the line dressing operations simply entailed having the chicken travel upside down between rows of butchers (generally women) who each performed small cuts on the bird, similar to the old meatpacking lines. Capital equipment needs were minimal—a longer chain operation, metal tables, knives, and sundry ancillary equipment. One study estimated that a $250,000 plant processing five thousand broilers daily in "New York dressed" form needed to invest just $25,000 more in equipment to add an evisceration department. But the labor needs were enormous. The same plant hired ninety-eight more employees to handle "line" evisceration operations, more than doubling its total paid labor force to 166. Most accounts indicate that this is when the labor force decisively shifted to African American women and away from white women who had better job options.[32]

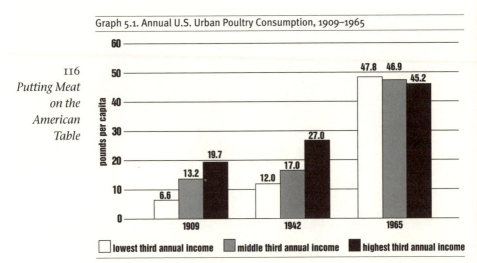

Graph 5.1. Annual U.S. Urban Poultry Consumption, 1909–1965

Sources: U.S. Department of Commerce, *Historical Statistics of the United States* (Washington, DC: Government Printing Office, 1975), 1:329–31; Great Britain Board of Trade, *Cost of Living in American Towns* (London: His Majesty's Stationery Office, 1911); U.S. Department of Agriculture, *Family Food Consumption in the United States, Spring 1942*, Miscellaneous Publication no. 550 (Washington, DC: U.S. Government Printing Office, 1942); U.S. Department of Agriculture Agricultural Research Service, *Food Consumption of Households in the United States, Spring 1965* (Washington, DC: U.S. Government Printing Office, 1966).
Note: Income levels correct for changes in consumer price index that doubled between 1942 and 1965. Lowest for 1942 is under $1,500; for 1965, under $3,000; for middle in 1942 is $1,500–3,000; for 1965, $3,000–$6,000; highest is over $3,000 in 1942 and over $6,000 for 1965.

With the widespread adoption of evisceration, the broiler industry consolidated its new place in American food consumption practices. Annual per capita poultry consumption tripled from its historic average of fifteen pounds per person to around forty-five pounds in 1965, while chicken's retail price fell more than 30 percent. A meat as expensive as round steak in 1928 cost one-third as much in 1965; at thirty-eight cents a pound, chicken was significantly cheaper than ham, pork chops, and hot dogs, and ten cents per pound less than hamburger meat. Chicken had lost its association as an expensive meal among American consumers; in fact, consumption was greatest at lower income levels, a reversal of historic patterns. The "Chicken of Tomorrow" had finally arrived.[33]

Chicken Becomes a Meat

With chicken consumption clearly on the rise, state extension services conducted more than a dozen consumer research surveys in the 1950s and early 1960s to assess how the industry could advance further. The studies showed that chicken had made great strides in consumers' eyes, reflecting the accomplishments of Gordy and his peers. Although consumers remained watchful for signs of bruising and poor bleeding, they made few complaints about chicken quality, testifying to the wide-spread adoption of the meat-type of broiler and improvements in processing technology. The bird had, in essence, been successfully standardized; it was a broiler, not a Plymouth Rock or Rhode Island Red. Declining retail prices, even given the added labor entailed in evisceration, indicated that the corporate integration, improved feeds, and concomitant production volume expansion had reduced costs significantly. Consumption patterns also showed that chicken was now a year-round meal with national appeal, and that the new self-service grocery stores were an asset to chicken sales, as broilers could be easily wrapped in clear film like cellophane.

All was not well with the new chicken, however. Consumers still considered chicken a special meal and preferred to serve it on Sunday as part of making that day distinct from the rest of the week. Housewives consistently ranked chicken third, after beef and pork, as a main course for conventional meals. Consumption also varied widely by ethnic background (Jews and African Americans were generally the best consumers), and lower-income consumers were more likely to treat chicken as an exceptional, occasional dish than higher-income groups. The expansion of chicken consumption had stretched but not yet altered its traditional place within American foodways.

"Chicken apparently has not fully achieved the status and prestige of a meat item," concluded one especially insightful study. "In fact, many housewives do not consider chicken to be a meat." While eaten more often than before 1945, chicken still functioned as a substitute for or alternative to meat. Before further advances in per capita consumption could be achieved, the study warned, it would be necessary to rid chicken of its "'weak sister' image and the 'inferiority complex' chicken has in relation to red meats." It recommended that industry expand consumption by launching "*an all-out attempt . . . to give chicken full status as a meat product.*"[34]

The studies also provided insights into why chicken was not yet a meat. Consumers explained they did not eat chicken more often because "it gets tiresome if eaten more than once a week." Such "chicken fatigue" reflected the relative monotony of chicken products available in 1960 compared to beef and pork. "Not only must she make a choice between beef, pork, and chicken," noted one study, "but she also must choose between beef steak and beef roast; pork chops and ham; whole fryers and fowl." These are, of course, not equivalent sets of choices. Beef and pork came in more varied cuts and flavors than chicken.[35]

A similar problem emerged from comments indicating the influence of family size on chicken consumption. An elderly woman explained she didn't buy chicken more often as "there's just my husband and I here now and it lasts too long with only two people." While two might have been too few to conveniently eat a three-pound broiler, large families also could be a problem. One woman explained she bought chicken only three or four times annually, even though her family liked it, because "with five kids, chicken just goes too fast." Unlike beef and pork, which yielded cuts of different size and cost, the broilers that dominated the early 1960s market were a "one size fits all" product. Leftovers from a three-pound bird were too much for a couple, and it was not cost-effective for larger families to eat chicken when they could instead obtain inexpensive beef and pork cuts.[36]

For chicken to become a meat, manufacturers would have to do more than redesign the bird and transform processing methods. They needed to reconfigure the form in which consumers encountered chicken in grocery stores and restaurants to create an array of options similar to beef and pork. Product differentiation and market segmentation, strategies that had served red meat well since Thomas De Voe and his colleagues had provisioned New York City, were now desperately needed for chicken to move out of third place as Americans' animal food of choice.

As poultry firms entered this new phase they could build on some very positive associations with chicken as a food item. Consumers of all types viewed chicken as a low-fat, healthy food that was easy to digest, hence good for adults and well suited for children. "I think of it as a way of pleasing my family," one woman commented. The absence of fat also meant that consumers saw chicken as an efficient, less wasteful food compared to pork and beef. "You can get more out of it for the money invested," reflected another housewife. Survey respondents also agreed that chicken was easy to prepare. "Chicken fits well into the concept of modern living," shrewdly observed a 1960 study. Capitalizing on these notions over the next two decades would catapult chicken consumption ahead of beef and pork.[37]

Two of the most successful—albeit quite different—product differentiation strategies were developed by Frank Perdue and the father-and-son duo of John and Don Tyson. Like Gustavus Swift a century before, Frank Perdue was remarkably knowledgeable of the animal that his firm transformed into food. John Tyson (father of Don) bore more similarity to Philip Armour because they shared a savvy knowledge of the marketplace. Through quite different paths the Perdue and Tyson firms would effectively promote higher chicken consumption levels and turn what once was a type of poultry into a form of meat.

John Tyson's initial entry into the poultry business gave his firm a commercial orientation from its inception. A Missouri produce buyer and trucker unfamiliar with the chicken industry, Tyson moved to Arkansas in 1931 to ply the hay market for animals in the drought-ravaged area. The growing broiler industry in the state's northwest corner attracted his interest, and in 1935 he expanded trucking operations to convey live chickens to Kansas City and St. Louis commission markets. Not until the early 1940s did Tyson actually enter the broiler business by starting a hatchery and a chicken growing operation, just in time to ride the enormous increase in Deep South chicken production following World War II. Anticipating opportunities distinguished Tyson's firm from its birth.

Don Tyson took over leadership of the firm in the 1950s and made it America's largest poultry company by creating new forms of, and hence markets for, its products. In 1964 Tyson sold the U.S. Armed Forces Commissary on precooked, portion-controlled chicken. A few years later he became the principal supplier of McDonalds' Chicken McNuggets and persuaded Burger King to sell Tyson chicken patties. For retail store purchases Tyson expanded into chicken hot dogs and precooked chicken products for home use. Because the white-meat dominated products created a surplus of dark meat, Tyson expanded his international export operations to regions of the world, especially Asia, where residents preferred chicken legs and thighs. His firm's broilers, instead of simply coming to the store whole, might end up in chicken hot dogs, military rations, and Japanese dinners.[38]

Tyson's market differentiation strategy for chicken echoed beef's appeal, especially the way different cuts and products at different price levels catered to distinct markets. The weakness of this beef-style approach, however, was that the chicken remained, well, just a chicken, with the firm that produced it hidden behind the commodity. Frank Perdue pursued a strategy more akin to early-twentieth-century pork producers who used branding to develop consumer loyalty to a particular firm's products and thus secure a reliable market share. While Tyson's success proved to post-1960 poultry producers that product differentiation promised a path to success, Perdue demonstrated that chickens could be branded similar to bacon and ham and hence removed from the status of a beef-style commodity food.

Branding entails close attention to product quality, a feature of the Perdue firm since its inception. The company began when Frank's father Arthur Perdue established a table egg farm in 1920 in the middle of the Delmarva peninsula. For twenty years the family firm remained a hatchery producing high-quality egg-producing birds. Similar to other Delmarva hatcheries Perdue Farms entered broiler production by contracting with local farmers to raise their chicks. Frank became company president in 1950 and augmented company operations by building mills to supply feed to its farmers. By 1968 Perdue was the largest broiler producer in the United States, selling 800,000 birds weekly to processors through the Eastern Shore Poultry Growers' Exchange.

When Perdue finally entered the processing business in the late

1960s, the father-and-son team had almost fifty years of chicken raising experience behind them. "I grew up having to know my business in
every detail," recalled Frank in 1973—similar to Gustavus Swift a century earlier. They also were latecomers to a highly competitive market
and under severe sales pressures with close to a million birds to sell
every week. Perdue set out to distinguish his chickens from the pack, to
in essence emulate successful pork producers like Oscar Mayer that had
used advertising to develop a following for their branded goods.[39]

The television advertising campaign launched in 1970 by Perdue in
the New York City area remains one of the most successful initiatives in
marketing history. The innovative advertisements featured Frank Perdue repeating, in many different situations, the company slogan, "It
takes a tough man to make a tender chicken." The notion that the firm's
president personally monitored the quality of its chickens was the theme
that permeated the campaign. Humorous advertisements also played
on chicken's perceived virtue as a low fat product. One depicted two
overweight customers eyeing the red meat freezer as Perdue cried out,
"Come on folks, shape up! Start eating my chickens." The advertisements successfully established Perdue broilers as a distinctive product
with highly positive results for the firm. In fifteen years from the com-
mencement of the advertising blitz, Perdue's output increased sixfold
as it rose from twelfth to fourth among America's poultry companies.[40]

Much of the branding strategy rested on the bird's allegedly superior
yellow hue. "My chickens always have that healthy, golden-yellow
color," Perdue declared in one advertisement. The color had little to do
with actual taste; instead, it was a visual indicator of quality for consumers who had favored yellow chickens since at least the turn of the
century. Perdue admitted he got the idea from Maine processors who
produced yellow chickens and "got a premium of three cents a pound."
He tried for the same advantage by adding xanthophyll to chicken feed
through natural ingredients such as marigold petal extract. Whether
there was any difference in taste is disputable, but that was not the point
or impact of Perdue's approach. He tapped into the way consumers rely
on visual signifiers to evaluate meat's quality. For consumers, the yellow in chickens was like the red in freshly cut beef, an ineffable feature
testifying, in some intrinsic way, to the product's wholesomeness and
value.[41]

Perdue also systematized combining chicken parts into assemblages
that appealed to different markets. Plant-packaged cut-up birds, once
known as "three-legged fryers" and "double-breasted chickens," accounted for less than 10 percent of the market in 1962. Perdue expanded
these types of products in the 1970s and 1980s priced to appeal to consumers at different economic levels, and packaged so that couples as
well as large families could obtain a convenient amount. Consumers

were able to buy packs of legs or thighs for low prices or breasts (sometimes boned and skinned) at higher levels. Other firms (including Tyson) emulated Perdue's branding strategy, so by the 1990s a consumer approaching a supermarket's meat department could find almost as many varieties among chicken as beef. Cut-up and "value added chicken" (including boneless parts, nuggets, hot dogs and patties) comprised 86 percent of chicken sales in 1995. That same year whole birds accounted for only 14 percent of the market, a long drop from 1962, when they accounted for 83 percent. During the same period, annual broiler production grew from 1.8 to 7 billion and per capita chicken consumption topped beef and pork. Through product diversification and branding, chicken had finally become a meat, and American's most popular one at that.[42]

Changing the form of chicken meat had drastic consequences on manufacturing technology and the chicken-processing work force. Plants became larger and production processes more complex, with industry employment doubling between 1975 and 1990, and then reaching 200,000 in 1995. Unions that had once held a foothold lost legal certification as older plants closed or changed hands. Yet despite the best efforts by firms to reduce their labor needs, in these so-called modern plants workers still stood shoulder to shoulder performing necessary tasks with their hands and small tools. The chicken remained an irregularly shaped natural product where processing operations had to be adapted to fit its particular anatomy.

While hand labor persisted in beef and pork killing eviscerating operations, automatic machinery displaced workers who had once performed those tasks in poultry plants. Establishing a detailed division of labor in the 1950s and 1960s facilitated mechanization, as inventive equipment suppliers and company engineers developed ways to automate repetitive tasks. In doing so, chicken processors had an advantage over their counterparts in the red meat industry: the smaller size of the animal permitted greater use of machinery in these initial stages than with the far larger and more irregular cattle and pigs.

The later stages of poultry production—segmenting the chicken into legs, thighs, wings and breasts, and then refining those pieces into boned and skinned parts—by contrast entailed adding labor-intensive cutting, boning, and processing operations. As with bacon, turning chicken into convenient packages for consumers made firms invest in massive new technologies and add tens of thousands of workers to their payrolls.

Chickens entered plants in bins unloaded from trucks, brought from farms where they were caught during the night. After men hung the chickens by their feet on a moving line, the birds were stunned by traveling through a shallow salt water bath charged with electricity. A deheading machine cut through the neck, allowing the chicken's blood to drain while the carcass remained on the moving chain. The killing room was saturated in red light because birds could not see that spectrum and hence were blind to what was coming. Vacuum fans designed to suck up dust and feathers filled the area with a loud din.

The bled carcasses went through a scalding bath and a series of "rubber finger" defeathering machines and then passed through a barrier of flames that scalded off remaining feathers. At this point the chickens' feet were cut off (later packed and shipped to Asian countries) and the chickens tumbled down a chute to workers laboring furiously to rehang them by their wings.

Once back on the moving chain, the headless chickens moved into the evisceration room, where a machine with twisting, piston-like plungers entered the animal's cavity and extracted intestines and internal organs. After a government inspector inspected the bird, a worker positioned the carcass so that a machine could cut off the intestines and organs. Another worker using her hands separated the hearts, livers, and gizzard into separate receptacles, later recombined and inserted (wrapped in paper) back into whole chickens. As the chicken continued along the chain, a worker used a suction device resembling a turkey baster to pull out the lungs and any remaining viscera before it went into the chill bath to cool for four hours with thousands of other carcasses. An evisceration department that once required dozens of workers was staffed by only a handful in the mid-1990s.

Once the body temperature reached 34° F in the chill bath, the carcass moved into the cutting operation. Inspectors diverted the best chickens directly into packaging, where they were wrapped and sold whole. For chickens destined to be cut into pieces for sale, a great deal more labor was required. As these are the forms of chicken meat that dominated the late-twentieth-century chicken market, the cutting and subsequent boning operations were generally post-1980 additions to poultry processing.

In the cutting room, mechanization took the form of automatic conveyer belts moving chickens from one work station to the next and machines that performed highly specific cutting operations. This process deskilled labor by allowing machines to assume actual cutting tasks. Nonetheless, the unique capacity of the human eye and hand to properly direct and position irregularly shaped chicken pieces made abundant hand labor still necessary for these operations. Machines severed carcasses into fore and aft sections, as well as into quarters and legs. Workers positioned the chicken for these operations by carefully placing it on the conveyer belt that fed into the machine performing the cut. As the parts fell back onto the conveyer they were whisked to more workers, who readied the pieces for the next cutting operation. Rudyard Kipling and other visitors to Chicago's huge 1890s slaughter-houses would have recognized the scene. As in the plants of a century before, workers stood shoulder to shoulder along the production line touching and moving animal meat every few feet, and at times trimming the carcass with hand-held knives to ensure that the machines performed their tasks properly.

The truly labor-intensive stages followed—boning and packaging chicken parts. Breasts went to the table boning operation, where women used sharp shears to cut the meat from bone and remove the skin. Chicken thighs moved into a boning room from the cutting line. Thigh boning machinery automatically pushed the bone out of the piece with a slowly moving piston, but to work properly depended on a worker inserting the thighs by hand into two dozen slots. The boned parts were then dumped onto a table surrounded by workers who used sharp scissors to remove fat and skin. Similar to the breast boning area workers labored closely together keenly observed by supervisors.

The boned and unboned parts met again along the packing line. Dozens of workers placed the pieces, brought there by conveyer belts or carried by hand trucks, into trays traveling along moving belts. Automatic machines weighed, wrapped, and labeled the trays before workers packed them into large containers to be moved by powered hand trucks into the dry chill area in preparation for shipping. Whenever possible, pricing the packaged chicken took place in the processing plant, so that supermarket clerks need not perform this task. Chicken companies that produced "house" brands for supermarkets labeled and priced items right there, if necessary creating separate streams for separate retail outlets, even though the chickens were the same.

A hundred thousand chickens per day flooded through this complex production process at line speeds of up to ninety birds per minute. The hanging, evisceration, and cutting jobs were especially relentless because the chickens arrived at work stations at such short intervals that there were few opportunities for breaks. A 1989 study reported that each "drawhand" along the evisceration line "pulled, twisted, and placed viscera of chickens in excess of 10,000 times per shift." The same report noted that deboners could handle up to 12,000 breasts in an eight-hour period. Handling chickens at such rates caused many abrasions and cuts, and too often produced repetitive motion injuries such as carpal tunnel syndrome. In 1989 chicken processing plants ranked second in the nation in repetitive motion injuries—behind only the red meat industry.[43]

While firms successfully applied technology to automate chicken killing, defeathering, and eviscerating, the need for hand labor remained staggeringly high in the latter stages of production. "If you see one person processing four pounds on the eviscerating line," explained an industry spokesperson, "you'll see four people processing that same four pounds in further processing." The increasingly important deboning operations remained the province of relatively skilled workers who used sharp hand tools rather than machines. Despite impressive advances in mechanization of cutting and packaging operations, the modern poultry processing plant bore little resemblance to the highly automated "flow" operations in hot dog manufacturing.[44]

Poultry firms had severe difficulties finding workers for these new jobs because of their low pay and rural locations that limited access to urban labor markets. Consequently, the poultry workforce's composition changed drastically in the late 1980s as immigrants, largely from Mexico and Central America, streamed into the industry.

The workforce's transformation was closely connected to the new production requirements. These onerous jobs, paying about 60 percent of the average wage for American manufacturing since the mid-1960s, drew workers at the bottom rungs of the American labor market. Regardless of the personal inclinations of employers, the highly competitive industry and labor-intensive process made paying wages higher than the industry average a prescription for bankruptcy. A high wage strategy similar to that of the unionized mid-twentieth-century red meat industry was only feasible if all leading firms had to bear roughly similar labor costs. In the absence of unions commanding a majority of poultry workers and able to raise wages across the board, firms followed a low wage strategy and accepted turnover rates as high as 100 percent annually.

These dynamics meant that the African American workers who had comprised the industry's principal labor force in the 1960s were largely supplanted by the mid-1990s by immigrant labor. African Americans already working in poultry jobs were likely to remain (especially if they held better-paid jobs), but younger blacks generally looked elsewhere for employment. In part this reflected the superior options available in service jobs for African Americans who had better language and literacy skills than the immigrants. On the Delmarva peninsula, for example, young African Americans often preferred to enter the thriving coastal tourist industry, just a few miles from the interior chicken plants. But the demographic shift also reflected a widely held judgment among employers that the immigrants were better workers. "Our experience is that the Hispanics are very conscientious and grateful for nice jobs," explained one plant manager in the late 1980s. "We have problems with blacks," complained another, "30–40 percent do not care if they work or not." These comments inadvertently admitted that what a Guatemalan immigrant might consider a "nice job" simply was not appealing to young African Americans seeking a better future.[45]

High-volume processing posed other problems for poultry operators. Charges of contaminated chicken bedeviled the industry, as the news media and watchdog groups repeatedly found evidence of bacterial contamination in chickens. As early as the late 1960s an unpublished study by the Delaware Extension Service found salmonella contamination rates "as high as 90 to 100% of the dressed broiler carcasses on any given processing day." With almost cyclical frequency in the 1980s and 1990s, newspapers, magazines, and television shows ran exposés with titles like "Is Chicken Safe To Eat?" documenting high levels of bacterial contamination. Requirements adopted in the 1990s that firms include safe-handling guidelines on packaged chicken reflected an acceptance by government regulators and the industry that contamination was an endemic problem.[46]

Bacterial contamination stemmed from the very same production methods that had made chicken an inexpensive and popular meal. The small size of chicken meant that mass production processing methods tended to mingle the carcasses together and hence aid and abet contamination. Bacteria could most easily spread in the feather removal operations, which left a great deal of dust and feathers in the air that could move from one bird to the next, and in the chill bath, where one bird could infect the liquid mixture and contaminate other chickens.

Salmonella may well have been widespread in red meat processing, but the different nature of processing operations minimized its impact. Cattle and hogs are not intermingled during processing as much as chicken because of their greater size, and their hide (where contamination begins) is removed and not eaten with the meat, unlike chicken skin. While processors could turn chicken into a meat in the eyes of consumers, it remained relatively a very small animal, and hence susceptible to a different array of health dangers in processing operations.

Consumers seemed to respond to exposés of salmonella and other forms of bacterial contamination such as campylobacter by being more careful in their handling and cooking of chicken, rather than changing their eating habits. Despite the sensational language of these stories and the well-documented incidence of bacterial contamination of chicken, consumption continued to climb. Between 1990 and 1995, when stories on chicken contamination saturated the news media, per capita consumption increased almost 15 percent to sixty-nine pounds annually; moreover, the rate of increase was the same as between 1985 and 1990. Consumers may have noticed the stories, but the exposés had no statistical impact on the upward trend in poultry consumption. [47]

Lower contamination levels could be achieved by altering processing methods, slowing production speeds, or requiring defeathered chickens to be chilled the same way as pigs and cattle, hung separately in air-chilled rooms. These processing requirements would doubtless raise chicken prices and put smaller firms, who were less able to make those investments, out of business, but they would reduce the health hazards inherent in modern production methods. Though it has the power to issue these regulations, the federal government has been unwilling to do so, preferring instead to cede more authority to the firms and reduce the power of government inspectors. In the absence of federal requirements, changes of this type are unlikely. And while consumers might worry about salmonella and the wholesomeness of chickens, consumption levels indicate that they feel sufficiently confident to keep eating it.

Georgetown, Delaware, host of the 1948 "Del-Mar-Va Chicken of Tomorrow Festival," was a very different town fifty years later. Almost half its residents were recent immigrants, the largest group men and women of Indian descent who were legal refugees from the Guatemalan civil war. A huge Perdue plant on the town's edge employed over a thousand workers, more than 50 percent immigrant. Since J. Frank Gordy brought the festival to Georgetown the Delmarva industry had grown to employ close to 15,000 people, relying on chicken raised on more than 2,000 farms to produce over 12 million broilers each week. Even with these impressive numbers the region ranked only fifth nationally in poultry production at the end of the twentieth century.

The rise of the postwar poultry industry represented an unparalleled shift of American food consumption practices. For chicken to move from being a food for special occasions to an item eaten several times weekly took extraordinary initiatives to radically transform the animal itself and the way it was processed for consumption. As a result, chicken became more of a meat than a variant of poultry.

These changes took place all along the chicken-producing axis, from Arkansas through the lowland South and up to Delaware, from hatcheries to farms to processing plants. Close supervision of farming practices by the integrated firms and intense research by universities into feeding methods utterly transformed the tempo of chicken raising. In 1923 Mrs. Steele's chickens took 16 weeks to reach 2.2 pounds, and had a feed conversion ratio (the amount of feed needed to increase weight by one pound) of 4.7. In 1993 broilers took 6.5 weeks to reach 4.4 pounds, with a feed conversion ratio of 1.9.[48]

Once this "Chicken of Tomorrow" left the farm it entered an equally transformed business and processing environment. Displacing the congeries of small enterprises from hatchery to farmer to trucker to processor to commission agent were large integrated corporations internalizing these varied transactions. Replacing the crude feather-picking and cleaning of turn-of-the-century operations were factories interwoven with conveyer plants and bristling with machinery that struggled (not entirely successfully) to automate hand labor.

Despite consumers' worries about chicken's wholesomeness (and widespread concern over the industry's labor practices) the popularity of chicken meat rose steadily throughout the post–World War II era. Chicken's attractive price relative to other meat, already a factor in consumption by the 1960s, was even more pronounced thirty years later. At ninety cents per pound in 1990, whole chickens were less expensive than any other type of meat on the market, and fully 50 percent cheaper than ground chuck beef. Chicken breasts, while more expensive at $2.07 per pound, were still cheaper than hot dogs and stewing beef. Encouraged by these advantageous prices, and by the variety of chicken products available in grocery stores and eating establishments, chicken consumption reached sixty-three pounds per capita that year. By 2002, Americans ate eighty-one pounds of chicken annually, significantly more than their beef consumption. In the early twenty-first century chicken was America's favorite meat.[49]

Consumption practices eating out and at home changed appreciably along with these numbers. Commencing in the mid-1950s restaurant chains selling fried chicken displaced the local outlets of the 1930s that James Beard had decried. Kentucky Fried Chicken rose to prominence among the purveyors of fried chicken sold by the piece, becoming America's largest private food service organization for a brief time in the mid-1970s. In 1994 (as a unit of Pepsico), KFC sold the equivalent of eleven pieces of chicken for each American citizen.[50] The demands of fast food chicken concomitantly impacted production. The precisely timed frying vats demanded consistent sizing of the breasts and legs for reliable cooking, pressuring firms and the farmers that supplied them to turn out chickens at consistent weights and sizes. Some restaurant chains even pushed cooking back into production facilities, inadvertently producing disasters such as the Hamlet, North Carolina, fire that occurred when cooking fat ignited and two dozens workers died when they were unable to exit through illegally bolted fire doors.

The fried—and hence highly fattening—fast food market contrasted with home preparation that emphasized chicken's low-fat features. From Miss Leslie through James Beard chicken recipes had relied on flour, bread crumbs, cream, and other fatty ingredients to impart taste and body, and to retain the meat's flavors during cooking. Similarly, recipes into the 1970s assumed the chicken had been purchased whole; instructions on how to clean, dismember, and prepare the carcass frequently accompanied cooking information. Only in the 1980s did the winning recipes in the annual cooking competition sponsored by Delmarva Poultry Industry, Inc. (the inheritor of the Chicken of Tomorrow organization) begin specifying supermarket-purchased parts for ingredients. The winning 1996 recipe reflected the change in home chicken preparation practices. Chicken in a mushroom and cream sauce began by lightly browning boned and skinned chicken breasts in olive oil, and then adding ingredients such as low fat cream cheese and skimmed milk. An equivalent 1953 recipe called for first sautéing the chicken in butter and then adding heavy cream and cream of mushroom soup.[51]

Inexorably, and almost imperceptibly, chicken changed from poultry to a meat product. Its transition was so dramatic and successful that pork producers even sought to improve their sales by branding theirs as the "other white meat." It is a testimonial to the success of chicken promoters that twenty-first-century pork, rather than remaining a companion to beef as it had been through most of American history, aspired to be like chicken.

Chapter Six

Convenient Meat

In 1953, *Fortune* magazine hailed the "Fabulous Market for Food" in a postwar America, where rising incomes were being disproportionately directed into higher value-added products. The magazine drew particular attention to how those with rising incomes wanted "not only good food, but convenience built into the food as well."[1]

Fortune was more on target than its editors could realize. Convenience framed the postwar transformations in the meat (and other food) businesses following World War II. The meat Americans ate had to accommodate itself to the faster tempo of production and consumption, driven by firms seeking greater efficiencies, as well as by declining meal preparation times among families.

Shifts in family eating habits were at the core of the changes in consumer demand for meat. After 1950 the proportion of women in the labor force grew steadily, most notably among married women with children at home, a sharp change from prewar patterns. Not surprisingly, these working women sought meals that took less time to procure and prepare than the ones their grandmothers had cooked. The best estimates of changes in cooking time indicate that weekly food preparation absorbed twenty-eight hours of labor in 1943, declining to eighteen in 1968, then to eleven in 1987.[2] Facilitating the drop in meal preparation times was a plethora of new technologies and food products.

Women with families were not only working more; the structure of households also changed as the traditional nuclear family with two parents and children became a minority phenomenon. By the year 2000 nuclear families comprised only 23.5 percent of all households. They were outnumbered by both households with more than one adult and no children and single persons living without children. Even single parents with children comprised almost 10 percent of all households.[3] For these "nontraditional" families, convenience and flexibility in food preparation methods were highly desirable.

New household technology enabled changing food preparation habits. The mechanical home refrigerator with a usable freezer became a commonplace kitchen appliance in the 1950s. As late as 1940 many urban working-class families and rural households still relied on small iceboxes to keep their food from spoiling. These devices were not suited to keeping meat for long periods of time, especially in summer months. Reliable home refrigeration expanded opportunities for women to purchase convenient meats such as hot dogs and bacon that could keep for several days or even weeks. Microwaves played a lesser role, but their facility defrosting frozen items and heating leftovers added to the technological options for quick food preparation.

Eating out as a routine practice accompanied new home food consumption habits. Prior to World War II, consuming food away from home was a rarity for working-class families. In the 1909 food consumption survey, expenditures for eating out were so low they hardly registered on family spending records. By the 1930s workers might eat lunch near their factory or office, but going to a restaurant with children remained exceptional; two-thirds of a family's minimal eating out expenses were for meals at work. Habits changed in the 1950s. Restaurants were a welcome break for working women with children and reasonable dining options for small households without children. Fast food chains benefited from these demographic trends, as did countless family restaurants that offered a greater range of menu options than McDonalds and Kentucky Fried Chicken. By 1955, 20 percent of food expenses involved eating away from home; that proportion grew steadily to reach almost 50 percent in 2002.[4]

The changes in consumer demand for meat reverberated all the way back along the meat provisioning chain, influencing retailing, distribution, processing, and farming. Too much can be made of these postwar changes; while dramatic, they were of a piece with the efforts of the meat industry since the early nineteenth century to alter the natural character of their product. Nonetheless, shifting consumer demand emphasizing convenience accentuated the meat industry's preexisting struggle to tame nature. Overcoming the processes of decay and reshaping meat into easier-to-use forms necessitated greater interventions with chemicals and machinery. But in doing so, firms still found themselves unable to suppress meat's natural qualities, even in its highly processed, convenient forms.

Creating Convenient Animals

Creating convenient meat demanded changes on the farm among the animals that supplied it. For two centuries the cycles of birth and growth cradled the provisioning system as farmers, processors, and consumers adapted their activities to natural rhythm of the seasons and animal biology. Although the gestation period from fertilization to birth could not be changed, virtually every other aspect of animal growth and reproduction fell under the lens of scientific and technological experimentation.

Manipulation of animal biology sought to overcome the twin obstacles of natural growth rates and variations in size. Standardizing animals made meat content more consistent and facilitated mechanization of processing systems. Animal breeding for particular traits was not a new innovation, and postwar interventions into animal genetics differed only in degree from what had been practiced for centuries. But more dramatic innovations took place in the materials animals consumed for growth, and the conditions under which they were "grown" to reach desirable slaughtering weights.

The 1948 "Chicken of Tomorrow" contest heralded intensive efforts to change chicken biology. The development of the Vantress chicken, and the profits associated with sales of high-meat-yielding breeds, stimulated decades of research and innovation in chicken genetics. While none of these efforts crossed into the realm of true biotechnology, with experiments at the level of DNA, extensive work went toward cross-breeding that would produce consistent chickens.

These high-yield strains of chickens, though, were more fragile than the hardy barnyard varieties and needed expensive care and feeding. Reminiscent of the demands of high performance cars, which require special fuel and careful maintenance, the chickens of tomorrow were picky eaters and petulant residents of their abodes. At the same time as the Vantress birds were becoming the standard for meat-type chickens, poultry farmers began to experiment with new feeds and raising systems to maximize the biological potential of the new chicken.

Antibiotics figured prominently in postwar chicken-raising methods. During the war years penicillin had amazed Americans with its capacity to halt infection and save soldiers' lives. Afterward chemical firms looked for peacetime uses and experimented with animals to see if subtherapeutic doses could have beneficial influences on growth. In 1950, American Cyanamid showed that low levels of antibiotics placed in feed could increase weight gains among chickens by more than 10 percent. By 1953, the year the Delmarva Poultry Exchange opened, the FDA had approved adding the antibiotics penicillin and tetracycline to animal feeds.

In the new poultry formulas of the 1950s and 1960s, antibiotics joined mineral- and vitamin-enriched feeds based on soybean and corn extracts designed to maximize weight gain and ward off disease. Medication and vaccinations also could be delivered through feeds and drinking water. Such careful feeding methods stimulated changes in chicken-raising practices, as carefully blended food cocktails were expensive and best administered under controlled conditions. Guided by cooperative extension service representatives and feed company servicemen, farmers began to move their chickens indoors to environments where light, temperature, food, and water could be carefully controlled. Because farmers' income depended on efficient feed utilization and bringing their birds to desired weights as quickly as possible, they complied with the required investments in chicken housing. By the 1990s chicken houses could cost a half million dollars and were equipped with automatic feeding and watering systems, heaters that allowed them to be used year round, and powerful fans to keep them cool in the summer.

Corporate integration of poultry production was both cause and consequence of this system. On the Delmarva peninsula and in other southern poultry regions, feed became a source of competitive advantage, as local firms built their own feed mills, working with cooperative extension service staff and scientists from local colleges of agricultural sciences to create ever-improving mixtures and methods. Seeking consistent outlets for their feed, these companies developed contract-growing relationships with farmers, supplying chickens and food and also closely controlling how farmers would then raise the birds. Success rested not only with retail penetration by firms such as Perdue and Tyson but also through corporate integration that lowered costs on the chicken-raising side. Firms without branded retail lines could still thrive as providers of chickens sold as supermarket brands if they were sufficiently efficient chicken "growers."

The impact of technology and science on chicken production can be measured through the reduction of time to reach market weight and through feed utilization efficiency. Chickens came to market heavier, at an average weight of 4.7 pounds in 1995, up from 3 pounds in 1940. They also took only forty-seven days to head to the slaughterhouse, one-half as long as their 1940 ancestors. While the distinct contribution of antibiotics is hard to measure, an estimated 10.5 million pounds entered poultry feed in the late 1990s as feed utilization rates—the amount needed to produce an equivalent weight gain—dropped under two. Chickens of the 1990s gained one pound for every two pounds they ate, twice as efficient as fifty years before.[5]

Such short growing times in confined conditions allowed the poultry farmer to break the tie between the environment and chicken growth, as six or seven "harvests" were possible each year regardless of climate. Consistency in a chicken's genetics, feed, and raising conditions permitted careful control of the animal's growth so that when the chickens went to market they would be close to the same size. The spring chicken was now an archaic phrase, as it was always springtime inside the houses that could hold ten to twenty thousand chickens, eating precise feed mixtures under carefully monitored conditions, and producing birds of predictable size and weight.

Parallel changes took place among the institutions that raised pigs for slaughter. Hog rearing in the first half of the twentieth century remained closely tied to nature's rhythms, even though it was located in a commercial nexus rather than a family economy as in the days of Laura Ingalls Wilder's family. Formerly, farmers raised pigs as part of an integrated farm operation, feeding them leftovers from the home and inedible (to human) crops. Fall remained "butchering time," after the remnants of the grain harvest were fed to the hogs so they would reach peak weights for commercial slaughter. October through mid-winter was the "hog rush" in the meatpacking industry, as firms brought on additional workers to handle the massive influx of pigs that would be turned into hams and roasts for Thanksgiving, Christmas, and Easter. Farmers obliged by encouraging births in the depths of winter so that young pigs, ten or so months old, would be sufficiently fattened for slaughter by the next fall. Employment fluctuations in the meat industry mirrored the seasonal cycle in the 1930s; firms specializing in pork products could have twice as many employees in the winter as in the heat of the summer. Refrigeration may have liberated the work of slaughtering and processing pigs from the seasons, but the dynamics of the farm economy still tethered hog raising to nature.

Such casual methods were ill suited to the business of meat production following World War II. After 1950 agricultural innovations reduced hog raisings' close tie with nature, albeit less completely than with chickens. Antibiotics and improved feeds accelerated hog growth and shortened the time between hog birth and arrival at market weights. The greater "speed" antibiotics imparted was magnified by changing consumer preferences for leaner pork which encouraged producers to bring hogs to slaughter at lighter weights, generally slightly over 200 pounds. Antibiotics and improved feeds also allowed weaning of piglets from sows just days after birth, with the sow returned to breeding two months later. Through careful management, farmers could arrange for year-round production of pigs, and with feed supplied commercially by firms rather than from their own land, the association between slaughter and season gradually faded until the hog rush largely disappeared in meatpacking parlance and employment patterns.

As with chickens, commercial feeds and a shorter maturation cycle encouraged confinement agriculture. While the far larger size of hogs precluded individual farmers building structures holding thousands of pigs, various types of indoor facilities became increasingly widespread. Modern family-operated farms of the 1990s raised hogs in indoor spaces designed for particular growth stages. Prefarrowing areas were maintained for breeding and gestation, farrowing houses for births and care of piglets until weaning, and finishing pens to bring the pigs to a market weight. A far cry from the practices of nineteenth-century farmers who permitted pigs to wander in nearby woods, confinement methods spread after 1960, as they reduced losses due to disease and parasites and facilitated systematic and finely tuned feeding methods.

Nonetheless, so long as pig processing remained centered in the midwestern corn belt, the large number of individual farmers with long histories in the business precluded development of close integration between provisioning firms and agricultural operations. Large midwestern hog processors such as IBP and ConAgra bought most of their animals daily on the open market, much as packers had done since the heyday of nineteenth-century Cincinnati.

Hog raising partially converged with chicken industry practices in the 1990s as new firms entered pork processing outside the Midwest. One North Carolina–based firm, Smithfield Farms, established contract relationships with local farmers who had decided to abandon tobacco farming. Smithfield controlled production of piglets, moving those just weaned to special nursery areas to receive special diets that boosted growth rates. When the pigs were eight to ten weeks old and weighed between forty and sixty pounds, the company sent allotments to farmers to bring them to market weights. As in the poultry industry, these farmers never actually owned the animals and raised them with feed and under conditions dictated by Smithfield. In 1993, only 11 percent of hog sales to processors were through contracts and integrated operations. By 1999, 59 percent of purchases took place in this manner.[6]

These integration methods permitted corporate hog producers to exercise close controls over animal breeding and growth practices. Smithfield went so far as to acquire exclusive American rights to a lean British hog breed produced by the British National Pig Development Company. Closely controlling production and marketing, Smithfield developed a line of processed pork products under the Lean Generation label, emphasizing the nutritional benefits of pork from this particular breed.

Volatile hog prices mitigated against completion of the process of integration similar to the poultry industry, where 100 percent of the animals commercially slaughtered were produced under contract. One corporate confinement operation, Premium Standard, went bankrupt in 1996 when prices for finished hogs dropped below costs for the 80,000 pigs on its 37,000-acre Missouri "farm." Unlike the poultry industry, most large hog processors protected themselves from such risks by not actually owning the pigs raised on confinement farms, instead creating multiyear contractual relationships.[7]

Antibiotics and new feeding methods also changed the dynamics of cattle-raising agriculture, though without quite the level of integration in the chicken and pig sectors of the business. Prior to World War II, the heartland of cattle production rested in Midwest states such as Iowa, Illinois, Minnesota, and Kansas. Ranchers might choose to raise cattle from calves to market weight, or specialize in breeding stock, calving, or "feeding" cattle until they were ready for slaughter. Livestock auctions located in the stockyards around major meatpacking plants were the principal means through which processors obtained beef supplies.

Since the days of the first commercial meat markets, cattle raising had followed a rhythm of slow growth through feeding the animals on grass, followed in some cases by an intensive period of fattening on grain before slaughter to increase weight and improve the meat's quality. Increased beef consumption in the late nineteenth and early twentieth centuries stimulated systematic efforts to finish cattle on feedlots that specialized in the final stages of the cattle-raising operation. Grain-fed cattle fetched a premium in the retail markets because the soft, well-marbled meat meshed with consumer expectations of beef value. Private and public grading systems formalized the market's preference for grain-finished beef.

Increasing standardization of beef products was inhibited by the uncertainties inherent in bringing cattle to auction. Farmers had to balance the costs associated with extended fattening versus the increased price that might (or might not) be achieved at sale. Hence cattle of many different finishes and weights entered the stockyard auctions, with the highly trained buyers of the packing companies serving as arbiters of animal quality and value.

Through a complicated process commencing in the 1950s, producers of finished, corn-fed cattle developed more efficient feeds, cheaper grain inputs, and direct marketing relationships with packing firms. Many of these early feedlots were family operations that expanded the scope of their activities during the postwar beef consumption boom. In Kansas, for example, the feedlot operated by Earl Brookover started in 1951 with just 500 steers. By serving local farmers seeking a place to finish their animals, Brookover expanded to service 100,000 cattle annually in the mid-1970s. He bought cattle that had reached 650 to 700 pounds on grass, and fed them a carefully formulated grain-based diet that brought them to 1,000 pounds within six months. Packers purchased the animals directly without going through an auction, with most going to the new slaughterhouses around Garden City, Kansas, the nearest town.[8]

Brookover's operation, like many postwar feedlots, depended on a radical transformation of agriculture on the high plains of Kansas, Nebraska, and Texas. This arid region had ample grasslands, but corn and other grains were in short supply and expensive to ship from the corn belt. Beginning in the 1950s, farmers invested in immense irrigation systems for their fields that relied on a "central pivot" pipe that reached down long distances to an underwater aquifer. The pivot connected to a horizontal pipe suspended from large wheels that would drop water on the fields. Feedlots stimulated massive growth of grain production; in Kansas, fields under irrigation grew tenfold, from 200,000 acres in 1950 to more than two million in the early 1970s.[9]

In a process paralleling chicken and pig raising, feedlot operators sought to compress growing times and improve feed utilization efficiencies to increase profits from their animals. Antibiotics became an integral component of reformulated feed mixes. As feedlots became increasingly important in the cattle-finishing process, they incorporated the antibiotic Stilbestrol (popularly known as DES) into most grain-based feeds. Approximately 95 percent of animal feeders used DES by the early 1960s. Compared to cattle that simply ate grain, DES-fed animals achieved market weight five weeks earlier and consumed 500 pounds less feed. The government banned DES in the 1970s due to its potential to cause cancer, but operators simply replaced it with other antibiotic drugs. By 2000, feedlots (aided by antibiotics and high-protein feeds) could bring an animal to slaughter within a year and a half from birth, less than half as long as before World War II.[10]

A Congressional study covering the period April 1992 to March 1993 gives a good sense of feedlots' impact on beef cattle-raising practices. The report demonstrated that large meatpacking companies had developed close business relationships with feedlots, sometimes through direct acquisition, in order to guarantee consistent shipments of animals at predictable weights and finish. Large packing plants with daily slaughter in excess of 4,000 head purchased nearly half of their cattle from large feedlots less than 100 miles away. And just as these large packers dominated commercial beef sales, large feedlots dominated cattle sales. The largest 152 of 19,000 identified cattle sellers (feedlots, farmer-feeders, auctions, and dealers) each sold more than 32,000 head of cattle during the study period, and together accounted for 43 percent of all cattle sold in the United States.[11]

By the end of the twentieth century, meat producers' search for convenience had generated significant interventions into animal biology. Animals were "grown" more than they were raised—birthed, fed, and housed under conditions designed to maximize weight gain efficiencies, minimize feed usage, and speed the process from birth to death. Nature could be held farther and farther away due to mankind's efforts to remake livestock to supply food for a voracious market. Smaller animals such as chickens were easier to redesign than larger varieties. Pigs and cattle still came to packinghouses at different weights and sizes, frustrating efforts to mechanize production as completely as in the poultry industry. Yet the predictability of the animals' characteristics, and the steady supplies assured by the confinement and feedlot systems, interfaced effectively with producers' efforts to create convenient meats for consumers.

Packaging Convenience

Building convenience into the meat purchased by consumers was the ultimate objective of these alterations in animal biology. Standardizing animals, making delivery of meat more predictable and less seasonal, facilitated development of products in forms that were easier for consumers to use. The natural process of decay remained the principal obstacle for meat purveyors. Faster cures for pork or better processing equipment for re-created meats like hot dogs only speeded production; they did not solve the problem of shelf life, shrinkage, and spoiling.

Packaging became the new method purveyors used to attenuate meat's natural process of decay. By interposing a film between the meat and outside environment, packaging could create artificial conditions that, in essence, performed work once borne by curing or refrigeration. Rather than alter the meat, as in cured products, packages were highly mobile and relatively inexpensive methods for slowing the forces of nature. When combined with cures and refrigerants, packages could magnify the efficacy of these older means of extending meat's wholesome life.

Packages also altered the way consumers encountered meat. Most obviously, cuts could be sold of varying weights and composition, such as cut-up chicken parts or chopped hamburger meat. Doing so allowed meat packages to be customized for different market segments. Less popular products, such as internal organs, could be presented better for consumer sales. In all cases the artificial environment made meat appear more attractive by influencing color, smell, and presentation of a particular cut. Packages sold the meat, but not simply by presenting attractive cuts in forms that customers expected. Use of a film to separate the meat from the environment allowed for manipulation of its appearance, while seemingly neutral promotional information suggested possible culinary uses.

The development of cellophane packaging launched this trend in meat retailing. Produced in the United States beginning in the 1920s, cellophane was the first transparent packaging material available for consumer products. A cellulose (wood-based) product, and not a true synthetic, cellophane was adapted by the DuPont Company for many packaging uses in the growing self-service grocery stores of the late 1920s and early 1930s. It was initially a huge success with baked goods, but by the mid- to late 1930s DuPont targeted the meat business, especially consumer-oriented products such as bacon and sausage.

DuPont made systematic attempts to spread the cellophane gospel among meat consumers and manufacturers. The firm mounted large displays at the American Meat Institute annual conventions, promoting cellophane's ability to "clinch the sale" because of its eye appeal. In

doing so DuPont heavily promoted sliced bacon and other wrapped
meat products, drawing on market research surveys that showed dealers preferred packaged goods for their ease of handling and superior
display appeal. Its 1940 booth, which an internal report described as
"the largest in the Exhibit and dominating the show," stressed how cellophane "made possible" the major requirements of modern meat
merchandising:

Packaged Meats—for convenience
Brand Identification—for repeat sales
Visibility—to show the quality
Eye and appetite appeal—for impulse sales
Protection—for health[12]

Food manufacturers seized on cellophane to advance brand identi-fication among consumers. "The customer ordinarily sells herself more satisfactorily than any clerk can sell her," observed Armour's general sales manager in 1948.[13] Immediately after the war Armour had rolled out an extended line of 500 branded products with new packaging cre-ated by famed industrial designer Raymond Loewy. Motivated by the need for brand identity in the new self-service retail environment, the packages carried a distinctive logo—the Armour Star—on all packages, and minimal text to maximize the meat visible through the cello-phane. Since women "do about 90 percent of the food buying," noted an article on the new Armour packages, "an attempt was made to use colors known to have a feminine appeal." Even gossip columnist Hedda Hopper noticed Loewy's foray into meat packaging. "It looks like a cologne box rather than just another butter carton," she noted approvingly.[14]

Following the war DuPont sought to increase cellophane sales by boosting a revolutionary retail practice: self-serve fresh meat. The theme of convenience permeated DuPont's promotional efforts. Advertise-ments promised that cases filled with cellophane-wrapped meat would "Make shopping quicker, easier," end "waiting in line," permit selec-tion of "the weight or size you want," and provide "new menu ideas" by making economical cuts more easily available. Advertisements also stressed the value of being able to see the food directly, promising (not quite accurately), "Shows What It Protects! Protects What It Shows!" Cellophane empowered consumers, these advertisements implied, even as self-serve meats gave the food stores greater power to influence consumers' choices.

Self-service meat was an outgrowth of the emergence of self-serve stores of the early twentieth century where customers selected their own goods from open shelves. In the 1920s and 1930s there were many meat and grocery stores that combined self-service dry goods with a meat counter staffed by a cutter who waited on customers and promoted sales of particular cuts. The brown paper typically used to wrap meat was an impediment to self-service, as customers resisted buying meat "blind." The growing food chains were interested in improving productivity in their meat departments, as lines of shoppers jammed butcher counters at peak hours, but they were hampered by the materials at their disposal. Despite experiments by retailers such as

Cellophane Protects Food

Saves by keeping it clean...Saves by guarding freshness and flavor

SHOWS WHAT IT PROTECTS!

PROTECTS WHAT IT SHOWS!

DU PONT Cellophane

DU PONT · BETTER THINGS FOR BETTER LIVING ... THROUGH CHEMISTRY

A&P to develop self-serve operations, at the end of World War II there were virtually no self-service meat departments in America.

While it now seems simple for consumers to grab a wrapped steak from an open-front refrigerator case, making meat available in this form was highly inconvenient for food retailers. Self-serve may have been less work for consumers, but it loaded retailers with troubling labor and material needs.

"Prepackaging introduces problems," noted a *New York Times* article in 1948, "that are not so pressing in over the counter sales."[15] Creating the right kind of artificial environment for fresh meat was not easy. Overly effective packaging prevented meat from receiving sufficient oxygen to assume the red appearance consumers desired. But admitting too much oxygen inside the package quickly exhausted the myoglobulin at the meat's surface, resulting in an unappetizing brown color. Similarly, the packaging material had to retard moisture loss so that the meat did not dry out, but retaining too much water rendered packages musty and wet and could generate "off odors" due to bacterial growth.

These difficulties encouraged extensive technological research to manage the interface between demanding self-serve meats and choosy consumers. DuPont engineered an MSAD-80 cellophane variety with two different layers specifically for fresh meat. Placing one side directly in contact with fresh meat retarded discoloration and admitted sufficient oxygen to create the desirable red "bloom." Applying heat to the other side (which had different chemical properties) would seal it to itself. Using this material, meat cutters could preslice meat and pass it to wrappers (generally young women) who used a heated iron to seal the cellophane around the meat.

Effective wrapping materials did not end the problems faced by innovative retailers interested in creating self-serve meat departments. Keeping wrapped meat looking red and wholesome was difficult. DuPont warned that cellophane was not effective at temperatures well above freezing; meat that retained an attractive appearance for three days at 36° would only last a day and a half at 50°. And recognizing the retailers' practice of rewrapping and "freshening" brown meat, the company sternly lectured against using sodium bicarbonate or other agents "to maintain color," as doing so violated government regulations.[16]

Prepackaging created additional difficulties as well. More labor was necessary in the back room because wrappers had to be added to prepare cut meat for consumers. Sealing meat in packages was tedious hand labor; the cellophane sheets had to be properly cut, folded around the meat, then sealed with a hand-held hot iron. A 1949 USDA study showed that the typical food market with a self-serve department needed 200 additional hours of labor from wrappers each week—equivalent to five new workers! Moreover, increased sales did not reduce labor costs, as there were no economies of scale as volume rose. Some savings were possible among the highly paid meat cutters who no longer had to spend time wrapping meat. But at least in 1949, the net effect was a wash—what large stores saved in increased production by meat cutters they paid out to wrappers.[17]

Effective cellophane use depended on development of new open-front display cases. With Freon (also a DuPont product) displacing ammonia as the primary meat market refrigerant in the late 1940s, cases could be held under 40° F to keep the cellophane-wrapped meat fresh. Firms also experimented with using lights at different spectrums to reduce discoloration. And to handle the problems of messy cases due to shoppers sifting through packages for the best value, retailers had to add another new staff person whose sole job was arranging and re-stocking the packages to maintain an attractive display.

Even with these obstacles, the pull of consumer convenience was sufficient to shift meat retailing to self-service. From just 178 outlets in 1948, self-service stores reached 2,800 three years later and then crossed the 20,000 mark in 1958.[18] Moreover, these departments were generally located in the large supermarket chains that were taking the bulk of consumer spending on food, with their proportion of retail grocery sales rising from 28 percent in 1948 to 69 percent in 1963.[19]

Two decades after the war's end, shopping for meat had become a completely different experience. Rather than patronize small butcher shops located a short walk from home, Americans drove to large diversified food stores. Eighty-six percent of the respondents to a 1956 *McCall's* magazine study said they always used a car to go food shopping. And the size of these stores soared to an average annual sales of $225,000 by 1960, an increase of 150 percent just in the 1950s. The typical way to shop for meat in the early 1960s was at self-serve cases located in large supermarkets, a completely different practice than thirty years before.[20]

Discontent with packaging materials persisted, however, and prompted a second wave of innovation in the distribution and sale of fresh meats. Worried about low profit rates in the mid-1960s, meat producers and food retailers hired the management consulting firm McKinsey and Company to identify the principal problems with meat retailing. To explain the seeming paradox of high sales volume and low profit rates in the meat department, the consultants blamed the costs incurred by making meat ready for self-service. "Meat is substantially changed in form within the store," unlike other products that generally were "received and sold in the same form." The material and labor costs entailed by this effort simply soaked up most of the income generated by meat sales.[21]

Packaging materials contributed significantly to low profits. Back room labor costs remained high because cellophane was hard to automate—in 1966, 75 percent of self-serve meats were wrapped by hand. And cellophane just did not do a good enough job protecting fresh meat. It suffered under the hands of discerning shoppers who picked through packages for the best cut; McKinsey and Company found that 13 percent of the labor used in self-service departments was just to maintain the appearance of refrigerated cases. Its inherently stiff fibers prevented tight wraps around the meat, creating a "boxy" look and leaving packages susceptible to tears. "The sheer magnitude of these costs," the consultants argued, "puts an entirely new dimension on the need for better packaging materials." DuPont's cellophane, once midwife to self-serve meats, now was the chief perpetrator of the meat department's problems.[22]

By the mid-1960s, the meat industry began to look to new packaging materials, principally polyvinyl chlorides (PVC) made from coal and petroleum byproducts. Once commercially available, these compounds rapidly displaced cellophane as a packaging material. PVC films engineered for fresh meat had some of the same key features as DuPont's MSAD-80 cellophane: they could limit water vapor evaporation and admit specific levels of oxygen so that the meat would bloom, not dry out, and also not become soaked with moisture. But PVC had other properties that rendered it a superior wrapping material. Compared to cellophane it was far clearer, more resistant to punctures, and easier to use in mechanized wrapping operations. Its greatest advantage, however, was that it could be heated to shrink tightly around the meat. Not only did these properties make for a better appearance in the self-serve display case, airtight PVC shrink-wrapped around meat permitted the use of vacuum packaging, vastly extending shelf life.

A 1968 advertisement for "VisQueen PVC Film" captured these advantages. Featuring a picture of a strong-looking female worker humorously (if in poor taste) labeled "Jackie the Ripper," the advertisement described her as a gentle soul but one whose hands nonetheless managed "to rip any film they touch." This well-known problem of unnamed cellophane was solved by VisQueen film, which "stretches instead," allowing Jackie to use "less film per package" and making "tighter, neater packages" that "rarely, if ever" had to be redone. Wrapping was faster because VisQueen "clings as it touches" and "seals instantly." The manufacturer promised that in addition to turning "Jackie the ripper into Jackie the wrapper," it would make "purchasers out of passers-by" because of its "uniform, crystal-clear" appearance that locked in "meat's bloom, juices, and freshness." The manufacturers could not have fashioned a better sales appeal to satisfy McKinsey and Company's packaging recommendations. By the 1980s PVC and other synthetics had virtually eliminated cellophane from the marketplace.[23]

Vacuum packaging of fresh meat in PVC bags went so far as to change meat distribution between processors and retailers. These bags' durable characteristics, and the ease with which they could seal out the atmosphere, made them attractive to purveyors seeking new ways to convey meat from packinghouses to supermarkets. Beginning in the late 1950s, the Cryovac Company promoted use of PVC films to vacuum pack meat as the final stage in slaughterhouse operations. Stimulated in part by the growing portion-control cutting operations used for the hotel, restaurant, and institutional trade, packers experimented with boning and further processing beef in the packinghouse, sealing cuts in Cryovac bags, then quick chilling them. Eliminating atmosphere from the containers protected meat from the dehydration and discoloration that usually accompanied the freezing process. Brown while frozen, the meat bloomed rapidly once cut and placed in display cases, presenting customers with its reassuring bright red color.

The new containers allowed firms to do away with the century-old "swinging meat" method of shipping beef to butchers for preparation into consumer cuts. Rather than send a quarter out on a railroad car, firms such as Iowa Beef Packers (later known as IBP) broke the meat down into basic subprimal cuts such as ribs, loins, and rounds for direct shipment (often by truck) to supermarkets or distribution centers. This innovation quickly became known as "boxed beef," after the containers used to ship the meat. Once in the supermarkets, employees without traditional butchering skills could slice the meat into standardized pieces and wrap them in PVC film for display. With its cost advantage, boxed beef's rapid growth was phenomenal. In less than two decades it became the typical method for marketing beef. Sales more than tripled between 1971 and 1979 to 4.8 million pounds. By the late 1980s boxed beef's national market share exceeded 80 percent.[24]

Aside from the innovative packaging aspect of the operation, boxed beef required little new technology. Instead, it followed firms' traditional strategy to accentuate the division of labor by fragmenting previously skilled work into its constituent parts. Instead of the production line ending with the carcass split into four quarters, dozens of workers lined up along a moving conveyer belt and used knives to make minute, repetitive cuts in pieces of meat that went by, trimming fat, removing bone, separating one part from another. Although these workers still needed to know how to use a knife and make particular incisions, the repetitive character of the jobs made it possible to learn them in a few days or weeks. This new method of preparing beef replaced meat cutters with, in essence, minimally skilled factory workers by moving the stage of turning quarters into consumer-ready cuts from the retail store to the packinghouse.

By far the most important benefit of boxed beef to retailers was the reduced labor needs in the supermarket's meat department. Retailers also liked how boxed beef allowed them to adapt their meat department's selection to consumer demand by selecting more of one cut and less of another. An IGA owner in Connecticut praised its easy handling because "it's film wrapped." The meat did not dry out and there were fewer bacteria problems "because we don't open the packages until about ten minutes before processing into retail cuts." Convenience for the store owner finally had come to the self-service meat market.[25]

The same packaging innovation that created boxed beef also stimulated an elaborate secondary market for marginal meat products. Firms

producing boxed beef were left with large quantities of bones and meat scraps from parts of the carcass that could not be prepared for consumer cuts. Specialized meat jobbers entered the business of taking the bones and scraps and using advanced technology to process and sell the deboned meat to hot dog manufacturers and chopped meat distributors.
Known as "mechanical meat separators," the machines ground up bone and used pistons or other devices to press the mash against a screen that permitted muscle and fat through but halted (most) of the larger bone fragments. More advanced systems developed in the 1990s relied on hydraulic pressure, based on the principle that meat and bone have different flow properties due to their chemical composition. These "automatic meat recovery systems" used high pressure to literally extrude meat, separating it from the bone matter and permitting it to be combined with other recovered meats.

Although meat fragments had gone into sausage products for two centuries, vacuum packaging permitted specialized meat jobbers to systematically collect meat fragments. Firms such as Hudson Meats acted as meat brokers, collecting beef from many sources, using meat recovery systems to salvage meat from bones, then grinding up the mixtures and sealing the ground beef in vacuum-packed containers that could be sold to institutional food suppliers or retail stores. The Cryovac OS1000 film gave processors eighteen days to move unfrozen ground beef from their plants to retail outlets. Once at the stores, the burger meat could be opened, allowed to bloom, and wrapped for consumer sales.[26]

Packaging was integral to making meat "convenient" in the manner foretold by *Fortune* magazine in 1953, but with even more dramatic consequences than the magazine's writers had imagined. Packaging contributed to the transformation of meat processing, wholesaling, and retailing by changing the economics of distance and time, not just how the consumer purchased meat in the store. Married to improved postwar cures and the expansion of home refrigeration, the new packaged meat extended storage times and made it unnecessary to shop shortly before the meal for which the meat would be cooked. Approaching gleaming trays of fresh meat in display cases, holding cured bacon in their refrigerators for weeks before use, eating cheap hamburgers at the many fast food outlets of the late twentieth century, consumers had the convenient meat they desired.

The Perils of Convenience

The perils of such convenience, however, came under close scrutiny in tandem with its growth. Americans were slow to appreciate the means through which the meat industry was able to deliver meat so efficiently and cheaply; when they did learn, the resulting controversies roiled the systems so carefully created to keep America a meat-eating nation.

As early as 1950, the House of Representatives conducted hearings on the "Use of Chemicals in Food Production" in reaction to new ingredients in animal feeds and processed foods. Chaired by James Delancey (D-NY), the hearings opened a new phase of close government attention to the changes in the food industry. Out of these hearings would come, in 1958, the "Delancey clause" that banned food additives proven to cause cancer in people or animals. The artificial sweeteners known as cyclamates were the best-known casualty of this clause, but government investigators also targeted the nitrites in meat processing.

The food industry had begun using sodium nitrite to improve cured pork products such as ham and bacon in the 1910s, following the ban on borax and boracic acid. Its major contribution was accelerating the curing process and maintaining meat color, but nitrites also inhibited growth of botulism-causing bacteria. As consumer advocates began investigating the dangers of America's corporate food system in the late 1960s, they argued that nitrites, with their cancer-causing potential, should be banned in accordance with the Delancey clause.

As the controversy peaked in the 1970s, testing showed that high nitrite levels in bacon could leave residues of nitrosamines that had clear carcinogenic dangers. How the nitrites generated these residues was hard to determine, so the meat industry argued forcefully against such a ban. As with the controversy over borax and boracic acid, scientific knowledge simply did not give a clear answer, so the debate turned on the level of risk that government regulators should accept. While the industry vigorously denied there was anything wrong with the nitrites, they did not object strongly when the Food and Drug Administration established rules in 1978 that required reduction of nitrite levels in bacon and other cured meats. Consumer advocates did not achieve an outright ban, but they did force nitrite levels down and established the purview of the federal government to intercede in this area.

Antibiotic use in animals followed a similar trajectory of protest and regulation. Stilbestrol (DES), primarily used in cattle feed, came under intense scrutiny for its potential to cause cancer. Studies in the early 1970s confirmed DES residues in cattle after slaughter, and found that DES in very high quantities (far higher than levels found in beef cattle) was linked to cancer. This information caused Congress to ban DES under the Delancey clause. The evidence was far clearer than with nitrites; there was little doubt that DES could cause cancer and that residues remained in fed cattle. Moreover, there were other antibiotics and hormones available that could generate similar growth. Yet the industry fought the ban for years, afraid that it would spread to the other ingredients it was adding to the feed and processed meat. Congress finally banned DES in 1979, with no appreciable impact on meat prices or agricultural productivity. But antibiotics remained completely legal for animal-feeding purposes, their use regulated but in no way proscribed, despite persistent worries that overuse of antibiotics in livestock feed could generate resistant strains of bacteria.

Consumer protests also spread to consider the problems of fresh meat prepared in the new packaging-based manufacturing systems. Ground beef came under particular scrutiny because of a series of incidents in which the *E. coli* bacteria contaminated large batches of ground beef, causing widespread sickness and some deaths. Public awareness of *E. coli* contamination can be traced to a 1992 incident in which sales of contaminated hamburgers at a Jack in the Box restaurant killed four children and sickened hundreds. As cooking meat to 170° F kills *E. coli* bacteria, only those eating underdone meat felt its consequences. In the Jack in the Box incident, the patrons falling ill were those unfortunate enough to obtain hamburgers cooked on the edge of the grill, where the heat was insufficient to kill the pathogens.

Chopped meat production methods magnified the danger of massive *E. coli* outbreaks. The bacteria contaminated meat in the processing stages; if the affected carcasses were destined for cooking as steaks and roasts, the impact was limited. Only those eating meat from the contaminated animal were potential victims, and with the bacteria kept on the outside of the meat, cooking generally sterilized solid meat cuts. But when these contaminated carcasses commingled with clean ones in the chopped meat manufacturing process, *E. coli* could spread widely. Once the bacteria infected the equipment in meat chopping and packaging facilities it could contaminate all the meat that touched the machinery. Freezing patties or sealing the meat in vacuum packages only kept the bacteria quiescent; once warmed to room temperatures and exposed to oxygen, *E. coli* growth was rapid.

Once alerted to the signs of *E. coli*–related sickness, the Centers for Disease Control and Prevention (CDC) established that there had been incidents well back into the 1980s. After the Jack in the Box case it closely monitored reports of foodborne diseases and ordered massive recalls of chopped meat when there were confirmed reports of illness caused by *E. coli* that could be traced to a particular batch. A 1997 recall of 25 million pounds of ground beef produced by Hudson Foods bankrupted the firm. Doubtless many Americans ate *E. coli*–contaminated meat safely because they cooked it well, but its persistence in the ground beef processing sector indicates that it was an endemic problem rooted in meat production methods, not an aberration.

Antibiotics, meat additives, and foodborne illnesses were only part of the reevaluation of convenience at the end of the twentieth century.

Critics also identified the impact of industrialized meat on farmers and workers who actually made the meat, as well as the troubles of the communities in which they lived. The growth of feedlots and their close relationship with large meatpacking firms squeezed the income of individual ranchers who produced the calves that entered this system. The families who raised chickens for the large poultry farms might have appreciated the protections from financial ruin that come with contracting arrangements, but chafed under the strict requirements for raising chickens that kept their incomes low. Hog farmers objected to the growth of corporate farms linked to processing companies, claiming they lowered the price received for finished hogs and thus forced family farms out of the marketplace.

The communities in which these animals were grown also exhibited ambivalence on the outcomes. Much of the meat industry takes place in rural areas with limited economic options, so the large operations—feedlots, confinement farms, manufacturing operations—were eagerly sought to produce jobs, tax revenues, and a stable agricultural economy. The shift of chicken production to the Deep South, hog raising to North Carolina, and the beef feedlot-slaughterhouse nexus to Kansas and Texas increased income and sparked economic growth in those regions, but they also imposed high costs. The sprawling agricultural complexes for raising cattle, pigs, and chicken have seriously impacted the environment because of high water usage and huge amounts of waste disposal. Hog farms' disposal systems holding millions of gallons of liquid waste in containment "lagoons" have been susceptible to breakage and contamination of adjacent woodlands and streams. The most egregious case took place in June 1995 when 25 million gallons of pig waste overflowed the containment lagoons of Oceanview Farms in North Carolina and contaminated fields and water systems for miles. But the beef feedlots and family-operated chicken farms also remained under close scrutiny for the impact of their disposal methods on the environment.

There are similarly mixed reactions to the employment brought to these communities by the meat industry. Meat-processing plants have been an important source of employment and tax revenues for communities without many alternatives. But the nature of meatpacking work has cast shadows on its reception. A competitive business still relying on hand labor, meat-processing firms emphasize high levels of productivity at relatively low wages, and preferably under nonunion conditions. Unions had until the 1980s created blue-collar, middle-class jobs in beef and pork plants, but the collapse of organized labor during the Reagan presidency allowed a sharp real decline in wages (approximately 30 percent) and deterioration of working conditions. Technology also replaced many of the highly skilled beef slaughtering and pork cutting jobs. While most packinghouse workers still labored with knives, they performed simplified operations compared to a generation before. Meatpacking and the poultry industries consistently rank first and second for injury rates among U.S. businesses, and their pay levels are well below the average for manufacturing industries. While the typical injuries are far less serious than in construction and mining, the high incidence of cuts, abrasions, and repetitive motion disorders generates a steady stream of damaged, and at times disabled, workers. Low pay stokes a system of high labor turnover, generally 60-80 percent annually, as workers seek to improve their conditions by moving from plant to plant; given the frequent layoffs too, most packinghouse workers are now part of the working poor of this country rather than its middle class.

In the low unemployment environment of the 1990s, firms had difficulties maintaining stable workforces; consequently, the meat industries became magnets for immigrants seeking a foothold in America. Primarily immigrants from Mexico and Central America, but also including Asians from Vietnam, Cambodia, and Laos, these workers eagerly took the jobs that U.S. citizens eschewed. These immigrants transformed the racial politics of small, mostly white communities such as Georgetown, Delaware; Marshalltown, Iowa; and Garden City, Kansas, and brought urban problems to previously sleepy, stable communities. The meat industries found themselves on the liberal side of national debates over immigration, favoring more lenient treatment of legal immigrants with the clear right to work, and also opposing crackdowns to deport alleged illegal immigrants working in their plants. Operation Vanguard, a highly touted federal initiative in the late 1990s to apprehend illegal immigrants in the Midwest, almost paralyzed the meatpacking industry as workers fled before rumored raids by the Immigration and Naturalization Service (INS). Protests by the industry and leading politicians in Iowa and adjacent states forced the INS to halt its raids.

Criticisms of the meat industry from so many sectors—farmers, workers, consumers, aggrieved communities—inevitably led to renewed complaints of a modern "meat trust." There was ample evidence of the industry's political power: acceptance of environmentally questionable development plans by state governments, the reluctance of federal regulators to require changes in processing methods that would reduce health hazards, the setbacks to unions who sought to organize meat industry workers. And, by any measure, a few firms dominated red meat—IBP, ConAgra, Excell/Cargill, and Smithfield—while Perdue and Tyson were the big players in the poultry sector. In 2002 the four leading companies in beef processing controlled 81 percent of the market; in pork, 59 percent; and in poultry, 50 percent. In 1937, by comparison, 78 percent of the meat products sold in the United States came from the

Big Four of Armour, Cudahy, Swift, and Wilson. The meat trust had returned, albeit with different players from earlier in the century.[27]

Historical comparisons tended to obscure the destruction and reconstruction of the meat industry that the convenience revolution had prompted. The growth of supermarkets and their self-serve meat departments in the 1940s and 1950s undermined the power of the old Chicago-based packing firms and permanently altered power relations between food business sectors. While the small prewar butcher shops had little choice but to patronize the meat wholesale outlets of the large packing companies and accede to their terms, supermarkets had far more market power. With their control over access to consumers, the centralized meat distribution and cutting warehouses of supermarket chains bypassed the packers' distribution system and broke the Big Four's market power. By 1962, the old Big Four's share of meat sales was down to 38 percent. By the early 1980s most of these companies had disappeared, unable to compete with IBP and other low-wage firms, their usable assets snapped up by new corporate organizations.[28]

The new oligopoly of meat companies that took shape in the late 1980s operated in a different business environment. Meat-processing firms had to contend with powerful supermarket and restaurant chains that sold their products to consumers. Creating consistently sized chicken parts, bacon conveniently packed in vacuum-sealed PVC containers, and reliably marbled beef cuts all reflected pressure from meat retailers that complicated the technology and organization of production methods. Consumer preferences had percolated up to the old Big Four through market research studies and feedback from butchers. The modern meatpacking oligopoly had to respond to the requirements of immense orders from a few outlets if they wanted consistent sales for their meat, whether retailers like Wal-Mart and Safeway or restaurant operations such as Kentucky Fried Chicken and McDonalds.

Competition for these critical retail outlets and consumer dollars among a few gargantuan players, rather than illicit cooperation, explains in large part the constant effort to reduce costs in all phases of meat production. Fighting for a limited number of lucrative contracts, meat firms were impelled to do whatever they could to keep their prices down. Going beyond the minimum requirements of government regulations thus was an unwise business decision, as allowing processing, labor, or supply costs to exceed competitors' would have damaging consequences. Branded products provided modest protection, as consumers would pay more for meat items of predictable quality (whether Oscar Mayer hot dogs or Perdue chickens), partially shielding some meat companies from the downward price pressures from retailers. Still, the competition among supermarket chains for sales of branded meat products still forced meat companies to keep their costs as low as possible.

Workers and farmers were the losers in this matrix of power rela-
tions. When farmers had brought their animals to auction markets for sale, competition among meat producers for quality animals gave livestock raisers market power. The railroad-based meat industry (dominant roughly for a hundred years between 1860 and 1960) brought animals to central stockyards where buyers from several plants sought their supplies. While collusion certainly took place, more often farmers had the opportunity for the market to set their animals' price. Contracting arrangements reduced farmers' options, as they were obligated to sell to a particular firm at a set price. What they may have gained in predictability they lost in autonomy and options should they not agree with the price offered for their animals or the terms under which they were to raise them.

The modern oligopoly brought most packinghouse workers back to conditions more reflective of the early twentieth century than the unionized era. The transformation of production methods associated with efforts to make convenient meat undermined established workplace relationships and weakened the unions that had established collective bargaining in the 1940s. By the 1990s labor organizations had little power in the meat-processing industry. Reducing costs in meatpacking meant, above all, cutting labor costs by keeping wages low and production speeds high. Without labor organizations to exert upward pressures on wages and to influence shop floor relations, workers had to accept companies' terms or go elsewhere. Similar to farmers, the pressures for convenient meat limited meatpacking workers' options.

The costs of convenience were many—additives to animal feed, ingredients in the food, pathogens distributed in the processing system, transformations of rural communities, and creation of a low-paid, heavily immigrant labor force. These disturbing elements of America's meat provisioning system occasioned repeated exposés, and sometimes tighter regulation, but no overall change in direction of greater convenience for consumers and food producers.

Consumers rarely showed much awareness of the complexity of placing conveniently wrapped meats in their supermarket coolers. Indeed, that was the purveyors' objective: to make it seem simple for shoppers to get what they wanted for their daily meals.

Convenience, however, rested on ever greater intervention into nature, as meat's habit of coming in irregular sizes and deteriorating quickly had to be attenuated as never before. Implicit in the very notion of convenience was using technology to help mankind claim victory over the organic, subduing animals and their parts to the imperatives of the human race. Altering animal biology and growth patterns, tinkering with forms of processed meat, adding chemicals to feeds, creating more automated production methods—all were elements of relentless efforts to turn nature's bounty into products that fit with the modern lifestyles of our civilization.

The array and diversity of the search for convenience should defeat explanations that assume that it was but a few institutions who led this trend. Indeed, many of the largest players fell by the wayside, whether the old meatpacking companies pushed aside by IBP and Tyson or suppliers such as DuPont whose once-revolutionary materials became obsolete. Consumer demand, shaped by seismic changes in family structure and workforce participation, established the new terrain for meat consumption practices. While consumers could not determine their product options, they could indicate what they preferred; firms that were slow to respond did not last long in the competitive food business.

Yet within these changes, the continuity of patterns of consumer demand for meat remained steady. Consumption studies after 1965 are far inferior to their predecessors, so we must rely on production data to appreciate the structure of demand at the end of the twentieth century. Per capita meat consumption (including poultry) in 2002 (measured by the amounts purchased by consumers) totaled 215.5 pounds—on par with the levels of the early 1960s and above the historic range of 150–200 pounds from the eighteenth through the mid-twentieth centuries. Beef and veal consumption had fallen from heights in the 1960s and 1970s (peaking at 97.4 pounds in 1976), but at 68 pounds per capita remained higher than in the early and mid-twentieth century. At 48.2 pounds per capita, pork was lower than it had been early in the twentieth century but consistent with levels since the mid-1950s when its new products resuscitated demand. Lamb remained a marginal meat, with slightly over one pound eaten per year. The major change in consumption pattern came from poultry products, with chicken providing eighty-one pounds to every American's annual diet and turkey fourteen pounds. The decline in beef consumption after 1976 had been more than compensated for with increases in poultry.[29]

As we entered the twenty-first century America remained firmly, if somewhat guiltily, a meat-eating nation. Indeed, for the last two generations we have been on a binge, bringing the amount we eat to more than nine ounces a day—far more than we need to benefit from meat's nutritional properties. Meat remains a sign of the good life, the American life, and a valued item in our diets, even as Americans remain skittish about the wholesomeness of our food system.

Persistent Nature

The ease with which we go to the supermarket refrigerator case and browse among many meat options for our meals conceals the enormous challenges tackled daily by meat producers. Animals continue to arrive at the slaughterhouse in irregular sizes; once they have been killed, any failure in the complex systems transporting meat to consumers can generate enormous losses due to bacterial contamination. Manipulating animal growth cycles is a massive challenge only incompletely realized by producers and dependent on substantial scientific and technological intervention. Mechanizing meat production remains highly uneven and lags well behind accomplishments in industries such as automobile manufacturing.

Albeit for different reasons, the meat industry and its critics prefer to emphasize how contemporary production methods have subdued natural processes. But looking at meat consumption and production in America over the past 250 years reveals that nature has been stubbornly persistent compared to the puny efforts of human civilization to deny its power.

The incomplete victory over nature by our contemporary provisioning system guarantees a steady stream of controversies over meat. Increased concentrations of livestock and commingling of carcasses in processing and disposal heighten the consequences of bacteria and animal diseases that once might have been confined to a single farm or factory. Animals fattened in confined settings offer a breeding ground for disease; technology, in the form of antibiotics and tight biosecurity measures, is the meat industry's response. Animals are killed and cut up in factories with high line speeds and under conditions where carcasses interact with one another, giving bacteria such as *E. coli,* salmonella, and campylobacter an opportunity to run rampant; more technology, in the form of microbiological inspection and irradiation is again our response. Remnants of the animals unusable for human food find their way into animal feed; though sterilized, these fragments create opportunities for new pathogens, such as mad cow disease, to spread. These diseases are, in essence, nature's unintended way of striking back at humankind's efforts to overwhelm the organic with technology.

The victory of convenience in meat production thus remains in uneasy tension with consumer anxieties over wholesomeness—that is, our worries over the persistence of unpredictable nature in our manufactured meat products. Food contamination scandals may have had little statistical impact on consumption practices, but our very fascination with the details admit the anxiety many still feel about the meat they eat.

Government oversight has been our band-aid solution to the unknowns of our food-provisioning system. Indeed, societies have handled the problems of ensuring wholesomeness by giving governments regulatory powers since at least ancient China. But there is a deeper, underlying problem that regulation, by existing, inadvertently admits: our hubris that we can conquer nature and turn its animals and plants into forms convenient to human beings.

The food industry and its critics fight, in essence, over the boundaries and rules of government regulation. These clashes are cyclical, repeated in a predictable dance of shock and outrage over a new practice, defense by the industry, and resolution with new elements of government supervision, the contours of which vary as much due to the political alignments of the day as with the determinations of scientists.

At its heart, resistance to regulation reflects assumptions that human beings can subdue, not just tame, nature. Regulation and inspection reflect an admission that we can't succeed in completely controlling nature, and those who live by denying nature's power are loath to concede their weakness. Repeatedly, though, our meat industry has had to submit, grudgingly, under great pressure; inexorably, more inspection, more regulation, wins. And why? Because the animal body refuses to die—disorderly nature lives on despite our best efforts to use inorganic technology to control it.

Regulation mystifies as well by implying that meat can shed its link with the ungainly animals and natural processes that brought it into existence. As much as the sanitary mylar-wrapped steak package conceals the thousand-pound steer, a portion of whose flesh is inside, regulation promises a fantasy of meat denuded of bacteria and dangers to consumers. Nature cannot be denied, however, so long as we remain organic beings who rely for sustenance on the bread of the earth and fruit of the vine.

Notes

Preface

1. Jeremy Rifkin, *Beyond Beef: The Rise and Fall of Cattle Culture* (New York: Dutton, 1992); Nick Fiddes, *Meat: A Natural Symbol* (New York: Routledge, 1991); Carol J. Adams, *The Sexual Politics of Meat: A Feminist-Vegetarian Critical Theory* (New York: Continuum, 1990).

Chapter 1. A Meat-Eating Nation

1. James F. Price and Bernard S. Schweigert, eds., *The Science of Meat and Meat Products*, 3rd ed. (Westport, CT: Food and Nutrition Press, 1987); Harold McGee, *On Food and Cooking; The Science and Lore of the Kitchen* (New York: Charles Scribner's Sons, 1984); Albert Levie, *The Meat Handbook* (Westport, CT: Avi Publishing Company, 1970); Albert E. Leach and Andrew Winton, *Food Inspection and Analysis* (New York: John Wiley & Sons, 1909); Jean Stewart, *Foods: Production, Marketing, Consumption* (New York: Prentice-Hall, 1938).

2. Jane Carson, *Colonial Virginia Cookery* (Williamsburg, VA: Colonial Williamsburg Foundation, 1985), 29.

3. Ibid., 60.

4. Ibid.; Miss Leslie, *New Receipts for Cooking* (Philadelphia: T. B. Peterson, 1852), 10-11.

5. Quotes from Priscilla Joan Brewer, "Home Fires: Cultural Responses to the Introduction of the Cookstove, 1815-1900" (Ph.D. diss., Brown University, 1987), 164, 122-23.

6. Mary Arnold, *The Century Cook Book* (New York, 1899), 156, 146.

7. Ibid., 72-73.

8. Brewer, "Home Fires," 145-46.

9. Detroit Stove Company, *Cooking with Gas* (Detroit, n.d.); Helen Armstrong, *Gas Range Cookery* (Eclipse Gas Range, n.d.); both from imprints collection, Hagley Museum and Library, Wilmington, DE (HML).

10. Katherine A. Fisher, *Good Housekeeping's Book of Good Meals and How to Prepare Them* (New York: Good Housekeeping, 1927), 212.

11. Ibid.; Sarah Field Splint, *The Art of Cooking and Serving* (Cincinnati: Procter and Gamble, 1928).

12. Sarah F. McMahon, "'All Things in Their Proper Season': Seasonal Rhythms of Diet in Nineteenth-Century New England," *Agricultural History* 63, no. 2 (spring 1989): 130-51, and "A Comfortable Subsistence: The Changing Composition of Diet in Rural New England, 1620-1840," *William and Mary Quarterly*, 3rd ser., 42, no. 1 (January 1985): 26-65.

13. Sam Bowers Hilliard, *Hog Meat and Hoecake: Food Supply in the Old South, 1840-1860* (Carbondale: Southern Illinois University Press, 1972), 104-5; Richard Osborn Cummings, *The American and His Food* (Chicago: University of Chicago Press, 1941), 15.

14. Great Britain Board of Trade, *Cost of Living in American Towns: Report of an Enquiry into Working-Class Rents, Housing, and Retail Prices, Together with the Rates of Wages in Certain Occupations in the Principal Industrial Towns of the*

United States of America (London: His Majesty's Stationery Office, 1911). The survey covered 8,000 households in towns and cities throughout the United States. Using monetary conversion charts and correcting for variations in household and sample size, I have converted the study's findings into per capita data by income levels.

15. Louise Bolard More, *Wage-Earners' Budgets* (New York: Holt, 1907), 205, 217–19.

16. Hilliard, *Hog Meat and Hoecake,* 41.

17. Hazel K. Stiebling et al., "Family Food Consumption and Dietary Levels: Five Regions," U.S. Department of Agriculture Miscellaneous Publication no. 405 (1941), esp. 189–93.

18. Hazel K. Stiebling and Esther Phipard, "Diets of Families of Employed Wage Earners and Clerical Workers in Cities," U.S. Department of Agriculture Circular no. 507 (1939), esp. 123–24.

19. U.S. Department of Agriculture, "Food Consumption of Urban Families in the United States, Spring 1948" (1949), esp. 15–19.

20. U.S. Department of Agriculture Agricultural Research Service, *Food Consumption of Households in the United States* (Spring 1965).

Chapter 2. Beef

1. Thomas F. De Voe, *The Market Assistant* (New York, 1867), 418.

2. Anonymous, *America and the Americans* (London, 1833), 86.

3. Richard Osborn Cummings, *The American and His Food: A History of Food Habits in the United States* (Chicago: University of Chicago Press, 1940), 264.

4. De Voe, *Market Assistant,* 37, 54.

5. Ibid., 57, 87, 58.

6. Andrew Beers, *Farmers' Almanac and Register* (Kingston, NY, 1805).

7. Thomas Butler Gunn, *The Physiology of New York Boarding-Houses* (New York, 1857), 230, 113, 174.

8. *America and the Americans,* 91; Thomas F. De Voe, *Abbatoirs* (Albany, 1866), 29.

9. *Report of the Sanitary Committee of the Board of Health on the Concentration and Regulation of the Business of Slaughtering Animals in the City of New York* (New York, 1874), 9, 21.

10. "Petition of Inhabitants of the Seventh Ward for a New Market," in *The New York City Artisan, 1789–1825,* Howard B. Rock ed. (Albany: SUNY Press, 1989), 36–37; Thomas F. De Voe, *The Market Book* (1862; reprint, New York: Augustus Kelly, 1970), 589.

11. De Voe, *The Market Book* 532, 533.

12. Edward Spann, *The New Metropolis: New York City, 1840–1857* (New York: Columbia University Press, 1981), 127, 129; *New York Times,* June 29, 1873, in Thomas De Voe papers, New-York Historical Society, New York, NY.

13. U.S. Department of the Treasury, *Statistics of the Foreign and Domestic Commerce of the United States* (Washington, DC, 1864); Pennsylvania Railroad Board of Directors Minute Books, 4:424, Pennsylvania Railroad archives, Hagley Museum and Library.

14. *National Provisioner,* August 20, 1921, 47.

15. *Harper's Weekly,* October 21, 1882, 663.

16. U.S. Senate, Committee on Finance, "Wholesale Prices, Wages, and Transportation," Report 1394, Part II, 52nd Congress, 2nd Session (Washington, DC, 1893), 25–26, 88–90.

17. U.S. Department of Commerce, *Thirteenth Census of the United States, Vol. X: Manufactures 1909* (Washington, DC: Government Printing Office, 1913), 346.

18. Mary Arnold, *The Century Cook Book* (New York, 1899), 155; Artemas Ward, *The Grocer's Encyclopedia* (New York: James Kempster, 1911), 58, 61.

19. Frank Rivers, *The Hotel Butcher, Garde Manger, and Carver* (Chicago: Hotel Monthly Press, 1935), 7.

20. Jessie W. Harris and Elizabeth V. Lacey, *Everyday Foods* (New York: Houghton Mifflin, 1927), 162–63, 155.

21. U.S. Department of Commerce, *Fifteenth Census of the United States, Vol. I: 1930, Distribution* (Washington, DC: Government Printing Office, 1933), 47, 59, 250; E. L. Rhoades, "Chain Meat Markets," in *Studies in the Packing Industry* (Chicago: University of Chicago, 1929), 3.

22. John H. Cover, *Neighborhood Distribution and Consumption of Meat in Pittsburgh* (1932; reprint, New York: Arno Press, 1976), 83; Kelsey Gardner and Lawrence Adams, *Consumer Habits and Preferences in the Purchase and Consumption of Meat,* USDA Bulletin No. 1443 (November 1926), 43–44.

23. Prices from Cover, *Neighborhood Distribution,* 161. E. L. Rhoades, "Advertising of Meats by Chain Grocery Companies," in *Studies in the Packing Industry* (Chicago: University of Chicago, 1930), 15.

24. Baker Ice Machine Company, "Retail Grocers and Meat Markets" (Omaha, [c. 1920]), 1.

25. Fairbanks, Morse & Co., "Mechanical Refrigeration" (New York, 1920); York Ice Machinery Company, "York Refrigeration for Retail Meat Markets" (York, PA, 1928); Baker Ice Machine Company, "Retail Grocers and Meat Markets" and "Dependable Refrigeration for Groceries and Meat Markets" (Omaha, 1922). All in National Museum of American History, Washington, DC.

26. E. L. Rhoades, "The Management of Chain Meat Markets," in *Studies in the Packing Industry,* 2, 7; Rhoades, "Chain Stores and the Independent Meat Retailers," in *Studies in the Packing Industry* (Chicago: University of Chicago, 1929), 15.

27. Evelyn G. Halliday and Isabel T. Noble, *Hows and Whys of Cooking* (Chicago: University of Chicago Press, 1928), 193.

28. F. W. Wilder, *The Modern Packing House* (Chicago: Nickerson and Collins, 1905), 97.

29. Henry Ford, *My Life and Work* (Garden City, NY: Garden City Publishing Co., 1922), 80–81.

30. Wilder, *Modern Packing House,* 76.

31. *National Provisioner,* October 22, 1921.

32. Frank Hlavacek, interview by Rick Halpern and Roger Horowitz, UPWA Oral History Project, State Historical Society of Wisconsin.

33. Lloyd Achenbach, interview by Rick Halpern and Roger Horowitz, UPWA Oral History Project; Wilder, *Modern Packing House,* 80–85.

34. Wilder, *Modern Packing House,* 97, 99.

35. Ibid., 116–18; Louis Tickal et al., interview by Rick Halpern and Roger Horowitz, UPWA Oral History Project.

36. Wilder, *Modern Packing House,* 114–15; Halliday and Noble, *Hows and Whys,* 215.

37. Rhoades, "Management of Meat Markets," 12.

Chapter 3. Pork

1. Laura Ingalls Wilder, *Little House in the Big Woods* (1932; New York: Harper-Collins, 1971), 19. My thanks to my sister, Diana Horowitz, for bringing this passage to my attention.

2. *Chambers's Information for the People,* 14th ed. (Philadelphia, 1854), 1:630.

3. Mary F. Henderson, *Practical Cooking and Dinner Giving* (New York, 1881), 160.

4. Sarah F. McMahon, "'All Things in Their Proper Season': Seasonal Rhythms of Diet in Nineteenth-Century New England," *Agricultural History* 63, no. 2 (spring 1989): 135–37, and "A Comfortable Subsistence: The Changing Composition of Diet in Rural New England, 1620–1840," *William and Mary Quarterly,* 3rd ser., 42, no. 1 (January 1985): 65.

5. Thomas F. De Voe, *The Market Assistant* (New York, 1867), 78.

6. *Chambers's Information for the People,* 15th ed. (Philadelphia, 1859), 1:150; Charles Flint, *Eighty Years' Progress of the United States* (New York, 1864), 67, 34; Margaret Walsh, *The Rise of the Midwestern Meat Packing Industry* (Lexington: University Press of Kentucky, 1982), 34.

7. Charles Cist, *Sketches and Statistics of Cincinnati in 1851* (Cincinnati, 1851), 278; Frederick Law Olmsted, *A Journey through Texas* (New York, 1857), 12.

8. Cist, *Cincinnati in 1851,* 281, 286; Matilda Charlotte Houston, *Herperos: or, Travels in the West* (London, 1850), 1:283.

9. W. G. Lyford, *The Western Address Directory* (Baltimore, 1837), 289.

10. Houston, *Herperos,* 284; *Chambers's Information,* 14th ed., 1:156; *Williams' City Directory* (Cincinnati, 1864), 106; Cist, *Cincinnati in 1851,* 229.

11. U.S. Department of the Interior, *Ninth Census of the United States, Vol. III: The Statistics of the Wealth and Industry of the United States* (Washington, DC, 1872), 649, 714.

12. U.S. Department of the Interior, *Report on the Manufactures of the United States at the Tenth Census* (Washington, DC, 1883), 393; *Scientific American,* November 7, 1891.

13. Rudyard Kipling, *From Sea to Sea: Letter of Travel* (New York, 1899), II:148, 153.

14. Louise Carroll Wade, *Chicago's Pride: The Stockyards, Packingtown, and Environs in the Nineteenth Century* (Urbana: University of Illinois Press, 1987), 105; U.S. Department of Commerce, *Twelfth Census of the United States, Vol. 9: Manufactures, Part III* (Washington, DC: Government Printing Office, 1902), 388.

15. Lyford, *Western Address Directory,* 280.

16. Olmsted, *A Journey through Texas,* 9.

17. Flint, *Eighty Years' Progress,* 65.

18. F. W. Wilder, *The Modern Packing House* (Chicago: Nickerson and Collins, 1905), 270–72.

19. Cist, *Cincinnati in 1851,* 282.

20. Flint, *Eighty Years' Progress,* 66.

21. Artemas Ward, *The Grocers' Hand-Book and Directory for 1883* (Philadelphia, 1883), 166–68.

22. Rudolf A. Clemen, *By-Products in the Packing Industry* (Chicago: University of Chicago Press, 1927), 254.

23. *Twelfth Census of the United States, Vol. 9: Manufactures, Part III,* 406. Wilder, *The Modern Packing House,* 296.

24. Wilder, *The Modern Packing House,* 249.

25. Louis F. Swift, *The Yankee of the Yards: The Biography of Gustavus Franklin Swift* (Chicago: A.W. Shaw, 1927), 104.

26. Frank Rivers, *The Hotel Butcher, Garde Manger, and Carver* (Chicago: Hotel Monthly Press, 1935), 47; *American Heritage Cookbook and Illustrated History of American Eating and Dining* (New York: Simon and Schuster, 1964), 504.

27. *Chambers's Information,* 14th ed., 1:630.

28. Poem "Bacon" from the Yakima Museum archives, Yakima, WA. My thanks to Daniel Levinson Wilk for sharing this with me.

29. Thomas Ashe, *Travels in America* (London, 1803), 241.

30. Quote from Frances Phipps, *Colonial Kitchens, Their Furnishings, and Their Gardens* (New York: Hawthorn Books, 1972), 38–39.

31. I.C.S. Reference Library, *Packing-House Industries* (Scranton: International Textbook Co., 1902), 38:1.

32. Wilder, *The Modern Packing House,* 320–21.

33. Ibid., 320.

34. *Douglas's Encyclopedia* (London: William Douglas and Sons Ltd., 1901), 93.

35. McArthur, Wirth, and Co., *Butchers, Packers, and Sausage Makers* (Syracuse, NY: 1900), Hagley Museum and Library, Wilmington, DE; American Meat Institute, "The Significant Sixty," *The National Provisioner,* January 26, 1954, II:60, 242.

36. Wilder, *The Modern Packing House,* 322.

37. *Douglas's Encyclopedia,* 93.

38. Harvey W. Wiley, *Foods and Their Adulteration* (Philadelphia: P. Blakiston's Son and Co., 1907), 37, 553.

39. Wilder, *The Modern Packing House,* 306–7; "The Significant Sixty," II:270

40. "The Significant Sixty," II:271, 216, insert pp.2–3.

41. Jesse Vaughn, interview by Rick Halpern and Roger Horowitz, UPWA Oral History Project, State Historical Society of Wisconsin.

42. Advertisement in N.W. Ayer Advertising Agency Records, series 2, 43:2, Archives Center, National Museum of American History (NMAH); Vaughn interview; B. Heller and Company, *Modern Curing Practice* (Chicago, 1958), 9.

43. Griffith Research Laboratories, *Prague Powder* (Chicago, 1952), 3; B. Heller and Company, *Modern Curing Practice,* 5.

44. Vaughn interview.

45. Advertisements in N.W. Ayer Advertising Agency Records, series 2,43:4.

46. Mechanical Manufacturing Company, catalog (Chicago, 1916), 104; Mechanical Manufacturing Company, *Modern Machinery and Equipment for the Packing Plant* (Chicago, 1928), 242–43; Allbright-Nell Co., *Anco Sausage, Ham, and Bacon Equipment* (1948), 34–37; all at NMAH library.

47. Allbright-Nell Co., *Anco Sausage, Ham, and Bacon Equipment,* Catalog No. 64A (The Inland Press, 1948), 33.

48. Cincinnati Boss Company, "Eat Meat Permeator: For Perfect Permeation," Bulletin P-52 (Cincinnati, n.d.); quotes from attached article from the *National Provisioner,* December, 6, 1952.

49. Clarence Wiesman, "Research and Development in Vacuum Packaging," in *New Potentials in Consumer Packaging,* Packaging Series no. 48 (New York: American Management Association, 1955), 25–26, Hagley Museum and Library, Wilmington, DE.

50. John M. Ramsbottam, "Flexible Meat-Packaging Materials and Their Selection," in *New Potentials in Consumer Packaging,* 15, 26.

51. "The Significant Sixty," II:278; Mary Elizabeth Pidgeon, *The Employment of Women in Slaughtering and Meat Packing* (Washington, DC: Government Printing Office, 1932), 7, 28–29, 40, 140–41.

52. Pidgeon, *The Employment of Women,* 140–41; Alma Herbst, *The Negro in the Slaughtering and Meatpacking Industry* (1932; reprint, New York: Arno, 1971), 77–78.

53. Roger Horowitz, "'Where Men Will Not Work': Gender, Power, Space, and the Sexual Division of Labor in America's Meatpacking Industry, 1890-1990," *Technology and Culture* 38, no. 1 (January 1997): 187-213.

54. U.S. Department of Commerce, *Historical Statistics of the United States* (Washington DC: Government Printing Office, 1975), I:213.

55. U.S. Department of Agriculture, *Household Food Consumption Survey, 1965–1966,* Reports 1 (United States), 2 (Northeast), and 4 (South).

56. U.S. Department of Commerce, *1963 Census of Manufactures* (Washington, DC: Government Printing Office, 1966), vol. II, part 1, 20A-14.

57. "The Significant Sixty," II:238.

58. Anglo-American Council on Productivity, *Meat Packaging and Processing* (New York and London, 1951), 14.

Chapter 4. Hot Dogs

1. "Cost-Cutting Production Methods at Packer's Fair Exhibit," *National Provisioner,* May 27, 1939.

2. Charles Elventon Nash, *The History of Augusta County, Including the Diary of Mrs. Martha Moore Ballard* (Augusta, ME: Charles E. Nash, 1904), 417, 397; Elizabeth Lea, *Domestic Cookery, Useful Receipts, and Hints to Young Housekeepers* (Baltimore, 1878), 169.

3. Lea, *Domestic Cookery,* 169.

4. Jane Warren, *The Great Economical Tea Co. Cook Book* (Buffalo, NY, n.d.), 28; Catherine Beecher, *Domestic Receipt Book,* 3rd ed. (New York, 1848), 33.

5. Lea, *Domestic Cookery,* 170.

6. Thomas F. De Voe, *The Market Assistant* (New York, 1867), 105.

7. Artemas Ward, *The Grocer's Encyclopedia* (New York: James Kempster, 1911), 553–59; Cincinnati Butchers' Supply Co. catalog (Cincinnati, n.d.), 107, 109.

8. B. Heller and Co., *Secrets of Meat Curing and Sausage Making,* 2nd ed. (Chicago, 1908), 120; F. W. Wilder, *The Modern Packing House* (Chicago: Nickerson and Collins, 1905), 385.

9. Ward, *Grocer's Encyclopedia,* 555.

10. Sam Coslow and Larry Spier, "At a Little Hot Dog Stand" (1939), in Sam DeVincent Collection of Illustrated American Sheet Music, box 542A, Archives Center, National Museum of American History, Washington, DC (NMAH).

11. Wilder, *The Modern Packing House,* 344.

12. U.S. Department of Commerce, *Fourteenth Census of the United States, Vol. X: Manufactures* (Washington, DC: 1923), 56; U.S. Department of Commerce, *Fifteenth Census of the United States, Vol. II: Manufactures* (Washington, DC: 1933), 176.

13. Mary Elizabeth Pidgeon, *The Employment of Women in Slaughtering and Meat Packing* (Washington, DC: Government Printing Office, 1932), 141–42.

14. Philip Weightman interview, by Rick Halpern and Roger Horowitz, UPWA Oral History Project, State Historical Society of Wisconsin.

15. Pidgeon, *Employment of Women,* 24, 23.

16. Cincinnati Butchers' Supply Co., *Boss Abattoir and Packing Plant Equipment,* Catalog 54 (Chicago, n.d.), 171, NMAH Library.

17. Kinyon Brothers, *Kinyon's Improved Meat Chopper* (Raritan, NJ, n.d.), in Warshaw Collection of Business Americana—Meat, box 6, NMAH Archives Center.

18. Allbright-Nell Co., *Anco Machinery and Equipment for the Meat Industry,* Catalog no. 10 (Chicago, 1923), 263, NMAH Library.

19. Cincinnati Butchers' Supply Company, *Boss Abattoir and Packing Plant Equipment,* 182.

20. Wilder, *The Modern Packing House,* 368–70.

21. Simple hand stuffers were commonplace in nineteenth-century butcher shops; they generally consisted of a cylinder terminating in a funnel at one end and powered by a hand-turned crank. A casing mounted on the narrow end of the funnel received the meat from the stuffer. This apparatus was a great improvement over the hand-stuffing methods of the country kitchen. One supplier boasted that it "develops immense power, at the same time it turns very easy; a small boy can run it"; S. Oppenheimer and Co., *Packers and Butchers Guide Book* (New York, n.d.), 45; Warshaw Collection—Meat, box 6.

22. Cincinnati Butchers' Supply Co., *"Boss" Packing House Equipment,* Catalog M-22 (Cincinnati, n.d.), 55; *Boss Abattoir and Packing Plant Equipment,* 477; Pidgeon, *Employment of Women,* 25.

23. Pidgeon, *Employment of Women,* 22-27, 141–42; Upton Sinclair, *The Jungle* (1905; reprint, Memphis: Peachtree Publishers, 1988), 120; U.S. Department of

Labor, *Wages and Hours of Labor in the Slaughtering and Meat-Packing Industry, 1929,* Bureau of Labor Statistics Bulletin no. 535 (Washington, DC, 1931), 36–39.

24. Quote from Bruce R. Fehn, "Striking Women: Gender, Race, and Class in the United Packinghouse Workers of America (UPWA), 1938–1968" (Ph. D. diss., University of Wisconsin–Madison, 1991), 37. American Meat Institute, "The Significant Sixty," *The National Provisioner,* January 26, 1954, II:276–78.

25. Tillie Olsen, *Yonnondio from the 1930s* (New York: Dell, 1974), 134.

26. Virginia Houston, interview by Rick Halpern and Roger Horowitz, UPWA Oral History Project, State Historical Society of Wisconsin; Pigeon, *Employment of Women,* 50–51.

27. Allbright-Nell Co., *Anco Casing Cleaning Equipment,* Catalog 67-A (1954), 1; Olsen, *Yonnondio,* 134.

28. Allbright-Nell Co., *Anco Casing Cleaning Equipment,* 10.

29. Rudolf A. Clemen, *By-Products in the Packing Industry* (Chicago: University of Chicago Press, 1927), 271, 273.

30. John H. Cover, *Neighborhood Distribution and Consumption of Meat in Pittsburgh* (Chicago: University of Chicago Press, 1932), 87; U.S. Department of Agriculture, "Family Food Consumption in the United States, Spring 1942," Miscellaneous Publication no. 555 (1944).

31. Oscar G. Mayer Jr., *Oscar Mayer and Co.: From Corner Store to National Processor* (Princeton: Newcomen Society, 1970), 19.

32. Curing time was relatively less of a problem in comparison to ham and bacon. Grinding the meat into very fine pieces of relatively similar size before placing it in a curing solution accelerated the process immensely. Chopped meat took only a few days to absorb salt and saltpeter, and just one or two days to cure once sodium nitrite use became commonplace in the 1930s.

33. Cincinnati Butchers' Supply Company, *Boss Abattoir and Packing Plant Equipment,* 183, 195.

34. "The Significant Sixty," II:16–17, 325.

35. Ibid., II: 63.

36. "Meat for the Multitudes," *National Provisioner,* July 4, 1981, II:110.

37. Ibid., II:104–9. Mayer, *Oscar Mayer and Co.,* 16.

38. U.S. Department of Agriculture, *Food Consumption of Households in the United States* (Washington, DC: 1956), 70.

39. Armour advertisements in N.W. Ayer Advertising Agency Records, oversized 48:4 and 46:4, NMAH Archives Center.

40. Ibid., 47:1–3.

41. U.S. Department of Commerce, *1963 Census of Manufactures* (Washington, DC: Government Printing Office, 1966), vol. 2, part 1, 20A-15; U. S. Department of Commerce, *Sixteenth Census of the United States: Manufactures, 1939* (Washington DC: Government Printing Office, 1942), vol. 2, part 1, 57.

42. Velma Otterman Schrader, interview by Rick Halpern and Roger Horowitz, UPWA Oral History Project.

43. U.S. Senate, *Hearings before the Select Committee on Nutrition and Human Needs, Part 13A: Nutrition and Private Industry,* 90th Congress, 2nd Session, and 91st Congress, 1st Session, July 15, 17, and 18, 1969 (Washington, DC: Government Printing Office, 1969), 3903.

44. Griffith Research Laboratories, *Prague Powder: Its Use in Modern Curing and Processing* (Chicago, 1962), 62, NMAH Library.

45. Koch Supplies, *Sausage and Chopped Meats* (Kansas City, 1954), 10; Griffith Research Laboratories, *Prague Powder,* 50.

46. Griffith Research Laboratories, *Prague Powder,* 62.

47. American Meat Institute Foundation, *The Science of Meat and Meat Products* (San Francisco: W. H. Freeman, 1960), 355.

48. National Hot Dog and Sausage Council, "The Hot Dog Fact Sheet" (Elmhurst, IL, 1993).

Chapter 5. Chicken

1. "The Chicken of Tomorrow," video, produced by Cooperative Extension Service, University of Delaware, 1948.

2. Karl C. Seeger, A. E. Tomhave, and H. L. Shrader, "The Results of the Chicken-of-Tomorrow 1948 National Contest," University of Delaware Agricultural Experiment Station Miscellaneous Publication no. 65 (July 1948).

3. U. S. Bureau of the Census, *Historical Statistics of the United States* (Washington, DC: Government Printing Office, 1975), I:331; U.S. Department of Commerce, *Statistical Abstract of the United States, 1930* (Washington, DC: Government Printing Office, 1930), 330.

4. J. Frank Gordy, "Broilers," in *American Poultry History* (Lafayette, IN: American Poultry Historical Society, 1974), 373; Lewis Wright, *The Practical Poultry Keeper* (New York, 1867), 51–54.

5. Miss Leslie, *Directions for Cookery* (Philadelphia, 1854), 140–47.

6. Miss Leslie, *New Receipts for Cooking* (Philadelphia, 1852), 88–91.

7. Mary Ronald, *The Century Cookbook* (New York, 1899), 35, 186–87.

8. Thomas F. De Voe, *The Market Assistant* (New York, 1867), 133.

9. Ibid., 134, 136.

10. Gordy, "Broilers," 373–75.

11. Swift and Co., "Swift's Premium Milk-Fed Chickens," Warshaw Collection of Business Americana—Meat, box 4, Archives Center, National Museum of American History, Washington, DC (NMAH); *Ladies' Home Journal,* November 1905.

12. Ferdinand Ellsworth Cary, ed., *The Complete Library of Universal Knowledge* (Chicago, 1904), 206–7.

13. Swift and Co., "Swift's Premium Milk-Fed Chicken."

14. Jean J. Stewart, *Foods: Production, Marketing, Consumption* (New York: Prentice-Hall, 1938), 488–89; Artemas Ward, *The Grocer's Encyclopedia* (New York: James Kempster, 1911), 504.

15. F. A. Buechel, *Wholesale Marketing of Live Poultry in New York City,* U.S. Department of Agriculture Technical Bulletin no. 107 (May 1929). On kosher killing methods, see Hugh Johnson, "The Broiler Industry in Delaware," University of Delaware Agricultural Experimental Station Bulletin no. 250 (October 1944), 46–47.

16. Temporary National Economic Committee, *Investigation of Concentration of Economic Power, Part 7: Hearings, March 9–11, May 1–2, 1939* (Washington, DC: Government Printing Office, 1939), 2867–71; R. O. Bausman, "An Economic Survey of the Broiler Industry in Delaware," University of Delaware Agricultural Experimental Station Bulletin no. 242 (March 1943), 51; Johnson, "Broiler Industry in Delaware," 45–46.

17. Barker Poultry Equipment Co., *Catalog no. 44* (Ottumwa, IA, 1946), NMAH Library.

18. Frank Rivers, *The Hotel Butcher, Garde Manger, and Carver* (Chicago: Hotel Monthly Press, 1935), 72.

19. James Beard, *Fowl and Game Cookery* (New York: M. Barrows, 1944), 1.

20. Marian Tracy, *Complete Chicken Cookery* (New York: Bobbs-Merrill, 1953), vii, ix.

21. Survey of extension activity in Delaware can be found in M. M. Daugherty, "Short History of the Broiler Industry," University of Delaware Agricultural Extension Service Pamphlet No. 15 (July 1944), 3–5.

22. W. T. McAllister, "An Appraisal of Marketing Problems in the Delmarva Broiler Area," 1954, Willard McAllister Papers, Hagley Museum and Library, Wilmington, DE.

23. Kimberly R. Sebold, "The Delmarva Poultry Industry and World War II: A Case in Wartime Economy," *Delaware History* 25, no. 3 (1993): 200–214.

24. W. T. McAllister and R. O. Bausman, "The Retail Marketing of Frying Chickens in Philadelphia," University of Delaware Agricultural Experimental Station Bulletin no. 275 (July 1948); W. T. McAllister et al., "Consumer Preference for Frying Chickens Studied," *American Egg and Poultry Review* (December 1950); McAllister, "Appraisal of Marketing Problems."

25. W. T. McAllister, "Mr. Poultryman: Marketing Is Your Business," University of Delaware Extension Service Bulletin no. 56 (August 1951).

26. Don Palmer, interview by Roger Horowitz, January 19, 1993, notes in possession of author; W. C. Evans and R. C. Smith, "The Daily Spread in Prices among Broiler Flocks Sold on the Eastern Shore Poultry Growers' Exchange," University of Delaware Agricultural Experimental Station Bulletin no. 330 (January 1960).

27. Evans and Smith, "The Daily Spread in Prices"; J. Frank Gordy, from "Annual Report, Extension Poultry Specialists," January 1, 1953–December 31, 1953, Cooperative Extension Service papers, University of Delaware archives; University of Delaware, School of Agriculture, "A Situation Report on the Delmarva Broiler Industry" (Newark, DE, n.d.), 15–23.

28. Interest rate estimates from R. O. Bausman, "An Economic Survey of the Broiler Industry in Delaware," University of Delaware Agricultural Experimental Station Bulletin no. 242 (March 1943), 49–50; Frank D. Hansing, "Financing the Production of Broilers in Lower Delaware," University of Delaware Agricultural Experimental Station Bulletin no. 322 (October 1957).

29. Willard T. McAllister, "A Plan of Action for the Delmarva Poultry Industry" (c. 1959) and "Opportunities and Possibilities for Improving the Economic Position of Delmarva's Broiler Firms" (c. 1963), McAllister Papers.

30. Edward Covell Jr., interview by Roger Horowitz, February 1, 1995.

31. Perdue's father had operated an egg hatchery since 1925 and had started contracting with growers to raise his chickens in 1950. Acquisition of a feed mill in 1958 completed the first phase of integrated broiler production. *Mid-Atlantic Poultry Farmer*, August 25, 1992; *Southern Exposure* 17, no. 2 (summer 1989): 15–16; Frank Gordy, *A Solid Foundation . . . The Life and Times of Arthur W. Perdue* (Salisbury, MD: Perdue, Inc., 1976).

32. George B. Rogers and Edwin T. Bardwell, "Marketing New England Poultry: 2. Economies of Scale in Chicken Processing," University of New Hampshire Agricultural Experiment Station Bulletin no. 459 (April 1959).

33. U. S. Department of Commerce, *Statistical Abstract of the United States, 1965* (Washington, DC: Government Printing Office, 1961), 367.

34. Richard Saunders, "Socio-Psycho-Economic Differences between High and Low Level Users of Chicken," Maine Agricultural Experiment Station, A.E. Progress Report no. 2 (November 1960). Emphasis in original.

35. R. C. Smith, "Factors Affecting Consumer Purchases of Frying Chickens," University of Delaware Agricultural Experimental Station Bulletin no. 298 (July 1953), 5.

36. Saunders, "Socio-Psycho-Economic Differences," 34.

37. Ibid., 32, 8.

38. Stephen F. Strausberg, *From Hills and Hollers: Rise of the Poultry Industry in Arkansas* (Fayetteville: Arkansas Agricultural Experiment Station, 1995); Douglas Frantz, "How Tyson Became the Chicken King," *New York Times,* August 28, 1994, sec. 3.

39. Christian McAdams, "Frank Perdue Is Chicken," *Esquire,* April 1973, 116.

40. Ibid., 114.

41. Thomas Whiteside, "C.E.O., TV," *New Yorker,* July 6, 1987, 39, 46.

42. Strausberg, *From Hills and Hollers,* 93; data from Delmarva Poultry Industry, Inc.

43. David Griffith, *Jones's Minimal: Low-Wage Labor in the United States* (Albany: State University of New York Press, 1993), 177.

44. Ibid., 102.

45. Ibid., 156.

46. "Relationship of Processing and Marketing Practices to the Incidence of Salmonella on Ready-to-Cook Broiler Chickens," August 24, 1970, Cooperative Extension Service Records, Box 104, University of Delaware Archives; "Ruling the Roost," *Southern Exposure,* Summer 1989:12–17. "Risky Business: Arkansas' Poultry Empire," *Arkansas Democrat,* April 21–25, 1991; "Chicken: How Safe?" *Atlanta Constitution,* May 26 and June 2, 1991.

47. U.S. Department of Agriculture statistics from http://www.ers.usda.gov/Data/.

48. William H. Williams, *Delmarva's Chicken Industry: 75 Years of Progress* (Georgetown, DE: Delmarva Poultry Industry, 1998), 73.

49. U.S. Department of Agriculture, "Food Consumption, Prices, and Expenditures, 1970–93," Statistical Bulletin no. 915 (1994); American Meat Institute, *Meatfacts 1991* (Washington, DC, 1992), 41–43. Data from U.S. Department of Agriculture at http://www.ers.usda.gov/Data/.

50. John A. Jakle and Keith A. Sculle, *Fast Food: Roadside Restaurants in the Automobile Age* (Baltimore: Johns Hopkins University Press, 1999), 221.

51. Delmarva Poultry Industry, Inc., *50th Anniversary Chicken Cookery* (Georgetown, DE, 1998), 14; Tracy, *Marian Tracy's Complete Chicken Cookery,* 73.

Chapter 6. Convenient Meat

1. *Fortune,* October 1953, 139.

2. Joann Vanek, "Time Spent in Housework," *Scientific American* 231 (November 1974): 119; Beth Anne Shelton, *Women, Men, and Time: Gender Differences in Paid Work, Housework, and Leisure* (Westport, CT: Greenwood Press, 1992), 83. For these sources and the argument in this paragraph I am indebted to Maurine Weiner Greenwald and her paper presented at the 1994 Organization of American Historians meeting, "Mealtime over Time: Food Consumption and the Demographic Revolution in American Women's Lives, 1950–90."

3. Hayden Stewart et al., "The Demand for Food Away from Home," U.S. Department of Agriculture Agricultural Economic Report no. 829 (January 2004), www.ers.usda.gov, 5.

4. U.S. Department of Labor, *How American Buying Habits Change* (Washington, DC: Government Printing Office, 1959), 112; Stewart et al., "The Demand for Food Away from Home," 1.

5. William Boyd, "Making Meat: Science, Technology, and American Poultry Production," *Technology and Culture* 42, no. 4 (October 2001): 631–64.

6. Steven W. Martinez, "Vertical Coordination in the Pork and Broiler Industries: Implications for Pork and Chicken Products," USDA Agricultural Economic Report no. 777 (Washington, DC: Government Printing Office, 1999), 1, 10.

7. Richard P. Horwitz, *Hog Ties: Pigs, Manure, and Mortality in American Culture* (New York: St. Martin's Press, 1998), 113–18, 132–33.

8. Charles L. Wood, *The Kansas Beef Industry* (Lawrence: Regents Press of Kansas, 1980), 291–93.

9. Ibid., 289.

10. Alan I Marcus, *Cancer from Beef: DES, Federal Food Regulation, and Consumer Confidence* (Baltimore: Johns Hopkins University Press, 1994), 1; Michael Pollan, "Power Steer," *New York Times Magazine,* March 31, 2002, 47.

11. U.S. Department of Agriculture, "Concentration in the Red Meat Packing Industry" (February 1996), www.usda.gov/gipsa/pubs/packers/conc-rpt.htm.

12. International Information Bulletins, November 8, 1940, DuPont Company Papers, 47:15, Hagley Museum and Library, Wilmington, DE (HML).

13. W.S. Shafer, "Prepackaged Self Serve Meats" (Armour and Co., 1948), 14, DuPont Company Papers, 49:18.

14. "Armour Star to Shine on 500 Packages," *Modern Packaging,* August 1945, Raymond Loewy Papers, HML.

15. Jane Nickerson, "Fresh Meat in Packages," *New York Times Magazine,* August 15, 1948.

16. *DuPont Technical Information Bulletin,* February 18, 1948, DuPont Company Papers, 49:15.

17. U.S. Department of Agriculture, "Retailing Packaged Meats" (Washington, DC, 1949), esp. 11, 14.

18. Jerry Lee Mautz, "A Discussion of the History and Development of the In-Store Mechandising and Packaging of Fresh Red Meat with Emphasis on the Effect of Polyvinyl Chloride Film on a Traditional Cellophane Market" (M.A. thesis, Michigan State University, 1966), 13.

19. James M. Mayo, *The American Grocery Store: The Business Evolution of an Architectural Space* (Westport, CT: Greenwood Press, 1993), 189.

20. Frank J. Charvat, *Supermarketing* (New York: Macmillan, 1961), 48; John P. Walsh, *Supermarkets Transformed* (New Brunswick, NJ: Rutgers University Press, 1993), 42.

21. McKinsey and Co., "Improving Profits in Marketing and Distribution of Meat" (1964), 27, HML Imprints Department.

22. For automation limitations, see Mautz, "In-Store Mechandising," 36. McKinsey and Co., "A Top Management Approach to Meat Merchandising" (1965), 26, HML Imprints Department.

23. "Jackie the Ripper" advertisement, *Progressive Grocer,* April 1968, 203.

24. U. S. Department of Labor, *Technology and Labor in Four Industries,* Bureau of Labor Statistics Bulletin no. 2104 (January 1982), 4; *National Provisioner,* June 20, 1987, 9.

25. Cryovac Division, W.R. Grace and Co., "1986 Retail Beef Study," 10.

26. *Packinging Digest,* December 2003; "There's Still Fun and Profit in Meats," *Progressive Grocer,* June 1975, 62.

27. Donald D. Stull and Michael J. Broadway, *Slaughterhouse Blues: The Meat and Poultry Industry in North America* (Belmont, CA: Wadsworth, 2004), 158; Robert M. Aduddell and Louis P. Cain, "The Consent Decree in the Meatpacking Industry, 1920–1956," *Business History Review* 55, no. 3 (autumn 1981): 363.

28. Mayo, *The American Grocery Store,* 189; Richard J. Arnould, "Changing Patterns of Concentration in American Meat Packing, 1880–1963," *Business History Review* 45, no. 1 (spring 1971), 33.

29. U.S. Department of Agriculture data, from www.ers.usda.gov/Data/.

Suggested Further Reading

Cronon, William. *Nature's Metropolis: Chicago and the Great West.* New York: W.W. Norton, 1991.

Cummings, Richard Osborn. *The American and His Food.* Chicago: University of Chicago Press, 1940.

De Voe, Thomas F. *The Market Book: A History of the Public Markets of the City of New York.* 1862. Reprint, New York: Augustus Kelly, 1970.

Giedion, Sigfried. *Mechanization Takes Command: A Contribution to Anonymous History.* New York: Oxford University Press, 1948.

Hilliard, Sam Bowers. *Hog Meat and Hoecake: Food Supply in the Old South, 1840–1860.* Carbondale: Southern Illinois University Press, 1972.

Horowitz, Roger. *"Negro and White, Unite and Fight!" A Social History of Industrial Unionism in Meatpacking, 1930-1990.* Urbana: University of Illinois Press, 1998.

Horwitz, Richard P. *Hog Ties: Pigs, Manure, and Mortality in American Culture.* New York: St. Martin's Press, 1998.

Jakle, John A. and Keith Sculle. *Fast Food: Roadside Restaurants in the Automobile Age.* Baltimore: Johns Hopkins University Press, 1999.

Levenstein, Harvey. *Paradox of Plenty: A Social History of Eating in Modern America.* New York: Oxford University Press, 1993.

———. *Revolution at the Table: The Transformation of the American Diet.* New York: Oxford University Press, 1988.

McGee, Harold. *On Food and Cooking: The Science and Lore of the Kitchen.* New York: Charles Scribner's Sons, 1984.

Schlosser, Eric. *Fast Food Nation: The Dark Side of the All-American Meal.* New York: Houghton Mifflin, 2001.

Smith, Page, and Charles Daniel. *The Chicken Book.* San Francisco: North Point Press, 1982.

Strasser, Susan. *Never Done: A History of American Housework.* New York: Pantheon, 1982.

Stull, Donald and Michael Broadway. *Slaughterhouse Blues: The Meat and Poultry Industry in North America.* Belmont, CA: Wadsworth, 2004.

Wade, Louise Carroll. *Chicago's Pride: The Stockyards, Packingtown, and Environs in the Nineteenth Century.* Chicago: University of Illinois Press, 1987.

Yeager, Mary. *Competition and Regulation: The Development of Oligopoly in the Meat Industry.* Greenwich, CT: Jai Press, 1981.

Index